NATIONAL DEFENSE RESEARCH INSTITUTE

Maintaining Arctic Cooperation with Russia

Planning for Regional Change in the Far North

Stephanie Pezard, Abbie Tingstad,
Kristin Van Abel, Scott Stephenson

Prepared for the Office of the Secretary of Defense for Policy
Approved for public release; distribution unlimited

For more information on this publication, visit www.rand.org/t/RR1731

Library of Congress Cataloging-in-Publication Data is available for this publication.
ISBN: 978-0-8330-9745-3

Published by the RAND Corporation, Santa Monica, Calif.

Cover: NASA/Operation Ice Bridge.

Preface

Despite a period of generally heightened tensions between Russia and the West, cooperation on Arctic affairs—particularly through the Arctic Council—has remained largely intact, with the exception of direct military-to-military cooperation in the region. This report examines potential transformations that could alter Russia's current cooperative stance there. It analyzes current security challenges in the Arctic with regard to climate and geography, economy, territorial claims, and military power, and suggests some ways in which these could undermine Arctic cooperation. It concludes with recommendations for the U.S. government to manage the risks to cooperation posed by these various factors.

This research should be of interest to the many organizations, inside and outside the U.S. government, that are Arctic stakeholders and are concerned with current and upcoming transformations in the Arctic, the likely impact of those transformations on the region's security, and Russia's role in that security.

Funding for this venture was provided by gifts from RAND supporters and income from operations. The research was conducted within the RAND National Security Research Division (NSRD) of the RAND Corporation. NSRD conducts research and analysis on defense and national security topics for the U.S. and allied defense, foreign policy, homeland security, and intelligence communities and foundations and other nongovernmental organizations that support defense and national security analysis.

For more information on the RAND National Security Research Division, see www.rand.org/nsrd/ or contact the director (contact information is provided on the web page).

Contents

Figures and Tables

Figures

Tables

Summary

To date, the Arctic has been widely viewed as stable and peaceful, with cooperation between Russia and other Arctic states remaining possible in spite of heightened geopolitical tensions. For example, the Arctic Council has endured as a forum for cooperative policy shaping, agreements have been signed and abided by, and nations—including Russia—have participated together in search-and-rescue exercises. This report examines the following research questions:

- What factors have contributed to maintaining the Arctic as an area of cooperation, even when tensions among Arctic states were rising in other regions such as Ukraine, the Baltics, and the Middle East?
- Can these factors sustain cooperation in the face of further dramatic changes that will likely take place in the Arctic?
- If cooperation is threatened by these changes, how might U.S. policy help mitigate the effects of these factors and contain tensions?

While there are many transformations at play in the Arctic, we selected and examined four—maritime access, resources, continental shelf claims, and Russian reaction to North Atlantic Treaty Organization (NATO) presence—that appear to have the potential to drive a dramatic shift in regional geopolitics from an emphasis on cooperation to escalation of tensions. When possible, the United States should take steps to reduce the risks that these transformations pose to Arctic cooperation, which represents a key objective of current U.S. Arctic policy.

This report is based on research of open-source literature; conversations with international experts on the Arctic and Russia; insights from a May 2016 roundtable with additional subject-matter experts from the U.S. government, think tanks, and universities; and use of a computer simulation for physical maritime access.

Russia's Approach in the Arctic: Between Buildup and Cooperation

Russia's actions and rhetoric with regard to the Arctic have alternated between inflammatory and conciliatory, creating some uncertainty regarding its intentions in the region. Russia has increased military presence in its High North, but not to Cold War levels. Russian policy in the Arctic has been mostly cooperative, and inflammatory speeches or events (such as the planting of a Russian flag on the seabed near the North Pole in 2007) may be best understood as aimed more at a domestic audience than an international one. Overall, Russia has benefited from its cooperative stance on Arctic issues for three main reasons: First, the difficulties of operating in such a rigorous environment make it inherently beneficial to collaborate; second, a number of key Arctic issues—oil spills, for instance—are transnational, therefore requiring collective responses; and third, economic development and investments benefit from a peaceful and cooperative environment—a factor of particular importance to Russia, which views the economic development of the Arctic as a key strategic objective.

Upcoming Transformations in the Arctic

While cooperation on Arctic issues has been successfully maintained between Russia and other Arctic nations—Canada, Denmark, Finland, Iceland, Norway, Sweden, and the United States—the region is already experiencing, or will likely experience, major transformations in the short to long terms that may alter Russia's incentives to cooperate. Four such transformations have the potential to upset current Arctic trends:

1. climate and geographical changes that radically modify maritime access
2. global interest in Arctic exploitation that drives competition for resources
3. legal decisions, specifically the upcoming recommendations by the United Nations (UN) Commission on the Limits of the Continental Shelf (CLCS) regarding the claims that Russia, Denmark, and Canada have submitted or will submit
4. NATO presence in the Arctic region that Russia might perceive as a military threat warranting a response in kind.

Climate and geographical changes that radically modify maritime access: Diminishing sea ice is the primary enabler for maritime access in the region. We used a previously developed geographic information system (GIS)–based model called the Arctic Transit Accessibility Model, which uses estimates of surface maritime accessibility based on projected sea ice distribution and thickness—as well as assumptions about vessel ice class—to assess the implications of a changing climate on access to the maritime Arctic region. In the future, maritime access will increase only during the summers, and the Arctic will remain a seasonally accessible area for all practical purposes. Nevertheless, even increasing seasonal access has important implications for Russia, which, for centuries, has been able to rely on thick, persistent sea ice to create a physical barrier along its northern shoreline. This barrier is diminishing, leading Russia to reconsider how to control its vast northern border for strategic and economic purposes. One instance in which Arctic cooperation could be threatened is if continued intense seasonal access changes draw substantial foreign presence along and around the Northern Sea Route. Russian ambitions to control this seaway have been widely documented and publicized. Not only would foreign presence fuel Russian concerns over sovereignty and potential attacks on its strategic and economic assets, more activity in general could lead to an increased risk of sparking unintended conflicts.

Global interest in Arctic exploitation that drives competition for resources: Better prospects for access to the Arctic have raised questions about whether "resource wars" might occur with the growth of international interest in exploiting the Arctic. Resources are a key factor shaping Russia's Arctic policy. Over the long term, Russia appears keen to develop its Arctic territory and increase its ability to bring resources, particularly hydrocarbons, to global markets. Potential for high global energy prices, along with the development of the necessary infrastructure and access to extraction technologies, will be instrumental in determining the magnitude of impact from this factor.

However, Russia is unlikely to discontinue cooperation with other Arctic states solely due to angst over resources. Russia's oil, natural gas, minerals, fish stocks, and other resources are not under any major threat. In addition, destabilizing the region could limit Russia's potential for benefiting from them. Other than the upcoming CLCS decision (which will be discussed next), there are no major territorial disputes between Russia and its Arctic neighbors in which there might be substantial resources at stake. No non-Arctic states appear poised to clash with Russia over resource control. Further, the difficulty of resource exploitation in this harsh, remote region alone is sufficient to severely hinder economic profitability in many cases,[1] let alone if a conflict were to put at risk personnel, ships, and infrastructure needed to support these activities.

Upcoming recommendations by the CLCS regarding the claims that Russia, Denmark, and Canada have submitted or will submit: The upcoming decisions by the CLCS on the claims set forward by several Arctic states regarding the limits of their continental shelf could upset the current order, should those decisions not support Russia's claims. In this scenario, Russia might choose to resubmit a claim with additional scientific evidence. In addition, or instead, Russia might interdict Danish and Canadian exploratory teams in the contested areas. This could have serious security implications because Denmark and Canada are NATO members. However, there is no concrete indication that the Alliance would intervene in this case.

[1] Some of the most significant oil and natural gas resources are geologically complicated to extract, which, in combination with volatile global energy markets, has made additional exploration and energy investment in the Arctic largely unfavorable in the near future.

Alternatively, Russia might receive a positive decision from the CLCS but then overreach by interdicting or limiting the transit of international vessels over its continental shelf.[2] Russia appears unlikely to make such a move, however, because contesting a decision based on the UN Convention on the Law of the Sea (UNCLOS) might open a "Pandora's box" whereby other decisions, some of them to Russia's advantage, could be contested by third parties. UNCLOS also ensures that most of the Arctic seabed can only be claimed by Arctic coastal states—a rule that Russia has no interest in undermining.

NATO presence in the Arctic region that Russia might perceive as a military threat warranting a response in kind: Russia could perceive itself as being under military threat in the Arctic if NATO decides to extend its presence in the region. One way this could happen is through heavier NATO involvement in the Arctic—whether through increased military presence of NATO members, or through a higher involvement of the Alliance as an organization in the region. NATO has an interest because five of its member states are Arctic nations, and Russia has denounced Alliance presence in its near abroad. Another plausible scenario is if Sweden and Finland choose to join NATO, which could trigger Russian response due to fear of encirclement. Russia has already warned that it would react negatively to such a decision by its Nordic neighbors. Domestic politics may play a critical role in how Russia reacts. While there is little evidence that shifts in public opinion have shaped Russian President Vladimir Putin's foreign policy so far, the Arctic is an important domestic issue in Russia before it is an international or diplomatic issue, suggesting that Russia's Arctic interests could be used as a nationalistic stake to shore up domestic support, particularly in times of political and economic difficulties.

[2] The CLCS decision does not imply a right to govern the water surface or water column above the continental shelf. However, it is plausible that Russia could use a decision in its favor as leverage for controlling maritime activity (e.g., by justifying increased maritime patrols). We consider this scenario only because Russia has previously demonstrated a willingness to disregard international law to serve its interests (e.g., in Crimea and—the United States might argue—with respect to control over the Northern Sea Route).

Conclusion and Policy Implications

Our first two research questions focus on the factors that have maintained the Arctic as an area of cooperation and the ability to sustain such cooperation in the face of dramatic changes that will likely take place in the Arctic. Our analysis produced five key findings.

1. **Russia's current militarization of its Arctic region does not, in itself, suggest increased potential for conflict, with the exception of accidental escalation.** Russia is still a long way from reestablishing Cold War levels of military presence in the Arctic, and is unlikely to use Arctic-based assets effectively in other, more likely, contingencies—for instance, in the Baltics.
2. **Russia's cooperative stance in the Arctic cannot be taken for granted.** Future behavior cannot be confidently anticipated on the basis of historical patterns, although the number of mechanisms (e.g., agreements, diplomatic organizations) through which Russia cooperates on Arctic affairs could make it difficult to abandon this stance in rapid fashion. Destabilizing the region would also limit Russia's potential to benefit from its Arctic resources, which its national priorities clearly indicate it wishes to do. Yet even economic factors will not necessarily steer Russia toward cooperation in the future, particularly if its ambitions for enhancing its energy sector through northern oil and gas reserves grow increasingly out of reach.
3. **Projected declines in sea ice suggest Russia will likely continue to militarize the Arctic, if only to protect its strategic assets and infrastructure in the region.** Russia's northern shore will be more exposed, increasing its perceived vulnerability to potential attacks. Increased maritime access overall will reduce Russia's ability to control Arctic shipping lanes or block them in the event of a conflict.
4. **While Russia has mostly benefited from UNCLOS decisions in the past, there would be nothing to stop it from ignoring or distorting UNCLOS recommendations if it judged such recommendations contrary to its interests.** It is worth noting

that the UNCLOS decision itself bears little risk of conflict, at least in the short term. The rights it would recognize would not lead to actual resource exploitation for years, possibly decades.

5. **Russia would likely feel threatened by an expansion of NATO's role in the Arctic.** The Kremlin has shown consistent hostility to increased support for NATO in Sweden and Finland, and to a larger NATO influence in the region, suggesting that keeping NATO at bay is a solid, and permanent, tenet of its Arctic policy.

Our third research question focuses on U.S. policy options that could help mitigate the effects of the factors outlined above and contain tensions. The fact that Russia's behavior in the Arctic could change from cooperative to conflictual and is difficult to foresee warrants close attention to the region on the part of the United States. As indicated in the 2013 U.S. Arctic Strategy (which includes "Enhance Arctic Domain Awareness" as an element of its first line of effort), monitoring of the region may require encouraging improvements in Arctic region domain awareness and access, through continuing and (as necessary) expanding funding for:

- mapping (including of underwater topography)
- vessels and aircraft that can operate in Arctic conditions[3]
- maintaining existing infrastructure and assets
- development of multipurpose ports and airstrips that can facilitate access[4]
- enhancing communications systems to promote a safe operating environment and help avoid unintended conflict
- further allocating intelligence, surveillance, and reconnaissance assets that can help increase the transparency of foreign Arctic activities to help prevent misunderstandings that can lead to conflict.

---

[3] Requirements will have to be developed to express specific needs.

[4] Specific needs will have to be identified first.

Unpredictability also suggests that special care should be taken to avoid accidental escalation of small-scale incidents. This can be done through supporting activities that bring the United States and Russia together on Arctic issues—for instance, through institutions (such as the Arctic Council, the Arctic Coast Guard Forum, and the International Maritime Organization), joint activities (such as safety and environmental exercises, collaborative scientific research) and information-sharing (for instance, data related to commercial shipping traffic). It could also be done by reducing Department of Defense barriers to participating in international Arctic activities that involve Russia when the focus is military support to civil authorities (such as search-and-rescue exercises). Another option would be to create a forum dedicated to security issues beyond the existing meetings of the Arctic Chiefs of Defense Staff.

Russia's increased vulnerability on its northern shore and sensitivity to an increased NATO presence in the Arctic region writ large also suggests that even limited incursions of the Alliance for such activities as routine exercises have the potential to fuel tensions when seen against the background of stronger support for NATO on the part of Sweden and Finland. While this does not mean that NATO should in any way halt its activities in the region, it suggests the necessity of striking a balance between ensuring that NATO has some capability and experience to support Arctic operations without establishing a presence of the Alliance in the region that would create tensions between Arctic nations, and particularly with Russia. This includes supporting measures designed to strengthen NATO's ability to conduct operations in cold-weather conditions and pursue efforts started at the 2014 Wales Summit and confirmed at the 2016 Warsaw Summit to adapt to the new threat environment.

Finally, the United States would be in a better position to pressure Russia to abide by its commitment to UNCLOS if it were a UNCLOS signatory itself—a step that is mentioned in the U.S. Arctic Strategy as an element of the third line of U.S. effort in the Arctic.[5]

---

[5] White House, *National Strategy for the Arctic Region*, Washington, D.C., May 10, 2013, p. 9.

While there are substantial barriers to fully addressing these policy implications because of political, budgetary, and other challenges, failing to prepare for these transformations might have serious implications for some key priorities of the United States, such as promoting freedom of navigation, ensuring the safety and environmental security of U.S. citizens living in the Arctic, and maintaining domain awareness in a region that could become both increasingly militarized and economically significant.

CHAPTER ONE

Introduction

The Arctic (Figure 1.1), which had been a key strategic region for the United States and the Soviet Union during the Cold War, lost this status almost overnight as the Soviet Union collapsed.[1] In the 1990s, the United States pulled substantial forces and capabilities from the region, leaving only a fraction of its strategic assets and bases, while Russia let most of its Arctic military infrastructure fall into disarray. Now, international interest has returned to this remote region because of the growing realization that its maritime geography is changing fast because of the melting sea ice cover,[2] and because of several widely publicized events. There have been suggestions that the Arctic may witness a "war for resources,"[3] following the 2007 planting of a Rus-

[1] On the strategic role of the Arctic during the Cold War, see, for instance, Rolf Tamnes and Sven G. Holtsmark, "The Geopolitics of the Arctic in Historical Perspective," in Rolf Tamnes and Kristine Offerdal, eds., *Geopolitics and Security in the Arctic: Regional Dynamics in a Global World*, Oxon, N.Y.: Routledge, 2014; and Klaus Dodds, "The Arctic: From Frozen Desert to Open Polar Sea?" in Daniel Moran and James A. Russell, eds., *Maritime Strategy and Global Order: Markets, Resources, Security*, Washington, D.C.: Georgetown University Press, 2016, pp. 154–158.

[2] While the entire planet is warming, the Arctic is doing so at a "faster-than-average" rate. Megan Scudellari, "An Unrecognizable Arctic," National Aeronautics and Space Administration, July 25, 2013.

[3] See, for instance, Scott G. Borgerson, "Arctic Meltdown: The Economic and Security Implications of Global Warming," *Foreign Affairs*, Vol. 87, No. 2, 2008; "New Cold War for Resources Looms in Arctic," *Moscow Times*, April 16, 2012; Marsha Walton, "Countries in Tug-of-War over Arctic Resources," CNN, January 2, 2009; Terry Macalister, "Climate Change Could Lead to Arctic Conflict, Warns Senior NATO Commander," *The Guardian*,

sian flag on the seabed at the North Pole,[4] the 2008 U.S. Geological Survey (USGS) assessment that the Arctic might hold very large hydrocarbon resources,[5] the competing claims of several Arctic nations to extend their rights to the Arctic seabed, and increased interest in the region on the part of non-Arctic nations, particularly China. The illegal annexation of Crimea by Russia in 2014 and the war in Eastern Ukraine, along with the rebuilding and modernizing of Russia's military assets in the Arctic, have only made concerns about the possibility of far northern conflicts more prominent. Russia's willingness to ignore international law in Ukraine and incur economic penalties raises questions about whether this fractious behavior will spread to other areas, such as the Baltic region and the Arctic.

Yet the Arctic is in most respects a particularly stable and peaceful region, where cooperation between Russia and other Arctic states—the United States, Canada, Denmark, Iceland, Norway, Sweden, and Finland[6]—has been maintained while geopolitical tensions have risen elsewhere. Even during the Cold War, the United States and Russia adopted a fairly conciliatory stance on Arctic issues, engaging with each other and agreeing on a number of treaties and regulations.[7] More recently, during the course of the Ukraine crisis, the United States, Russia, and their Arctic partners successfully established new regional institutions, such as the Arctic Economic Council and the Arctic Coast

October 11, 2010; or, more recently, Sohrab Ahmari, "The New Cold War's Arctic Front," *Wall Street Journal*, June 9, 2015.

[4] C. J. Chivers, "Russians Plant Flag on the Arctic Seabed," *New York Times*, August 3, 2007.

[5] These include an estimated "90 billion barrels of oil, 1,669 trillion cubic feet of natural gas, and 44 billion barrels of natural gas liquids." USGS, "Circum-Arctic Resource Appraisal: Estimates of Undiscovered Oil and Gas North of the Arctic Circle," Washington, D.C., USGS Fact Sheet 2008-3049, 2008.

[6] These eight states are the member countries of the Arctic Council. Only Canada, Denmark, Norway, Russia, and the United States are Arctic coastal states.

[7] On these points, see Stephanie Pezard and Abbie Tingstad, "Keep it Chill in the Arctic," *U.S. News and World Report*, commentary, April 27, 2016.

Figure 1.1
Map of the Arctic

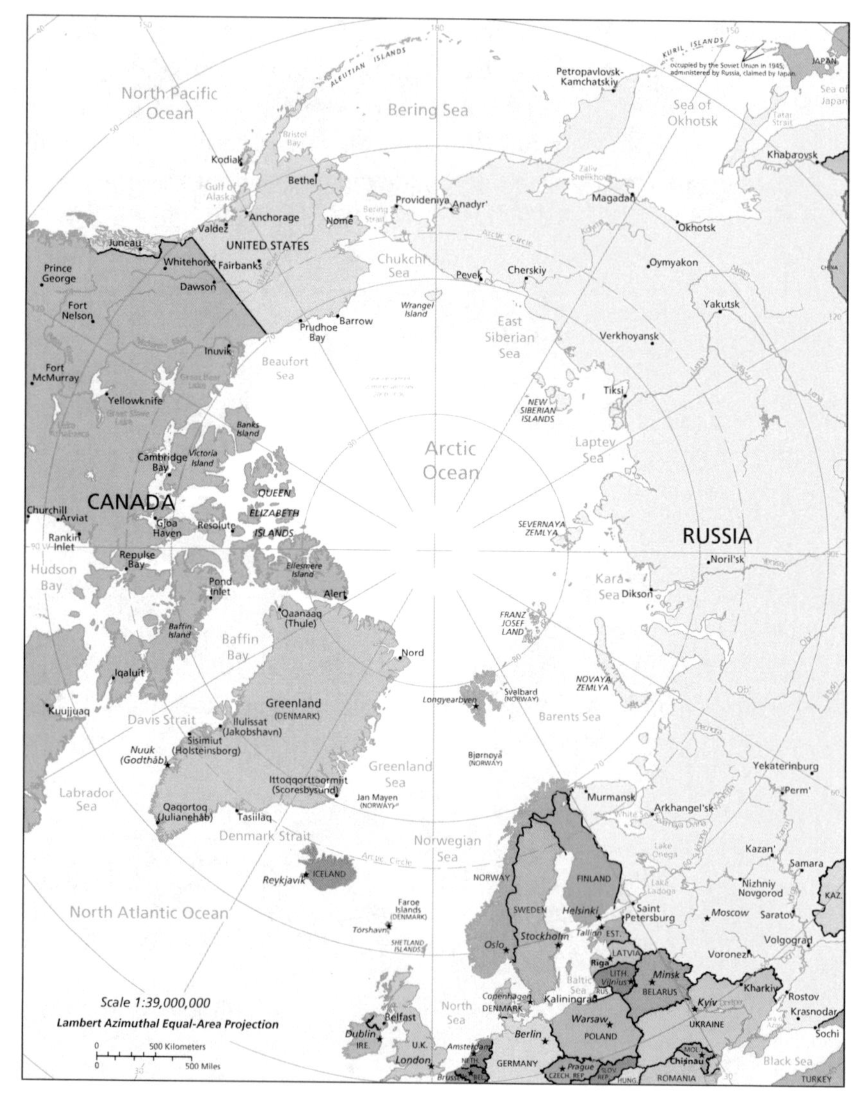

SOURCE: Central Intelligence Agency, Political Arctic Region," *World Factbook*, undated.

RAND *RR1731-1.1*

Guard Forum.[8] Overall, outside events have had limited spillover in the Arctic, although Russia and some of its Arctic neighbors have occasionally used the region as a loudspeaker for discontent with events elsewhere, perhaps precisely because actions taken in the Arctic have low risk of igniting conflict due to this stability. While a risk of spillover can never be entirely ruled out, the Arctic lacks some of the elements that make the Baltic region and Eastern Europe such contentious areas for Russia. Russia's neighbors in the Arctic are not former Soviet satellites that Russia considers its "Near Abroad" and where it intends to keep the influence it once had. Russia has taken exception to North Atlantic Treaty Organization (NATO) and European Union expansion pulling its influence away from Poland, the Baltic states, and threatening to do the same in Ukraine. In contrast, Russia's Arctic neighbors are part of the Western sphere, with four of them (the United States, Canada, Denmark, and Norway) being NATO founding members since 1949, and two others (Sweden and Finland) being officially neutral. These Arctic countries also lack substantial populations with linguistic and cultural ties to Russia. Such different strategic contexts make regional cooperation in the Arctic intrinsically easier, although it is not devoid of occasional tensions, as we will illustrate later. Finally, the Arctic environment itself, while growing marginally less formidable in the foreseeable future, should nonetheless continue to help deter conflict under most circumstances because of the presence of unforgiving operating conditions.

This paradox of a region that is effectively peaceful but routinely described as a potential future ground for conflict, and the questions it raises for U.S. policy, motivated this study. Maintaining a cooperative stance in the Arctic is one of the United States' three lines of effort in that region, according to its Arctic Strategy.[9] At the same time, the Arctic is undergoing major transformations that could change Russia's

---

[8] The concept of an Arctic Economic Council was adopted as early as May 2013 in the Kiruna Declaration, and the Council was approved in January 2014 (all prior to the Ukraine crisis), but the first meeting of the Council took place, with Russia as a participant, in September 2014. Arctic Council, "Arctic Economic Council," web page, September 1, 2015.

[9] White House, *National Strategy for the Arctic Region*, Washington, D.C., May 10, 2013, p. 2.

cost-benefit calculations when it comes to working with, or against, its fellow Arctic nations. This study thus examines the following research questions:

- What factors have contributed to maintaining the Arctic as an area of cooperation, even when tensions between Arctic states were rising in other regions?
- Can these factors sustain cooperation in the face of further dramatic changes that will likely take place in the Arctic?
- If cooperation is threatened by these factors, how might U.S. policy help mitigate the effects of these factors and contain tensions?

While there are many transformations at play in the Arctic, resource and time constraints limited the number of transformations we could examine in this study. As a result, we focused our attention on four of them: changed maritime access due to Arctic sea ice reduction; increased global interest in exploiting Arctic resources, along with new technological abilities to do so; the upcoming decision by the United Nations (UN) Commission on the Limits of the Continental Shelf (CLCS) regarding Arctic nations' claims over the disputed seabed; and increased support for, and involvement of, NATO in the Arctic region.

In selecting these factors, we examined broad areas of relevance to Arctic affairs—climate change, sovereignty, economy and resource extraction, security, social issues—and searched for major transformations that have already started showing effects in one or more of these areas, and fulfilled three criteria, including that they:

1. were identified as transformations that will create significant changes for both the United States and Russia based on a review of the relevant literature, 16 conversations with a total of 23 experts on the Arctic and Russia, including officials and think-tank analysts from four of the five Arctic coastal states (Canada, Denmark, Norway, and the United States), and discussions

from a RAND-organized roundtable event.[10] By "significant changes," we mean that these transformations will create new risks and new opportunities and will likely alter cost-benefit calculations of cooperation for Russia.

2. are already under way. In this sense, they are not imaginary scenarios but existing trends that have the potential, if or when they follow their course, to alter relations between Arctic nations significantly.
3. are core transformations from which further transformations will likely be derived. For instance, we could examine the impact on Arctic cooperation of a higher level of involvement of non-Arctic states (e.g., China) in Arctic affairs, but this factor is already encompassed in our first and second factors examining increased maritime access and global interest in exploiting Arctic resources.

Chapter Two examines our first research question by developing a theory for Russia's historical cooperation on Arctic affairs on the basis of observations from previously published work and conversations with subject-matter experts, in addition to inductive cost-benefit analysis. Chapter Three examines the four factors we identified in detail, synthesizing information from literature, conversations with Arctic experts, a roundtable event, and other supplementary information (e.g., from a maritime access simulation) to weigh whether and how each type of transformation is likely to alter Russia's cost-benefit calculation for Arctic cooperation in the near term. Finally, Chapter Four considers the policy implications of the findings in the previous two chapters and whether the United States should consider taking additional steps beyond what is outlined in its Arctic Strategy and associated Implementation Plan to help maintain the historical status quo of Arctic cooperation—which it has no intention (per strategy) of undermining—despite the effects the four factors may have on the region.

---

[10] The RAND Arctic Roundtable was organized to support this research and took place on May 4, 2016, in Arlington, Virginia, with 21 subject-matter experts from the U.S. government, as well as U.S. think tanks and universities.

CHAPTER TWO

Russia's Arctic Strategy: Military Buildup and Political Cooperation

This chapter provides some background on Russia's military and foreign policy in the Arctic region, with a particular focus on Russia's renewed interest in building up its military capabilities in the Arctic. This chapter also explores Russia's track record in cooperating with other Arctic nations, as a way of identifying the factors that have facilitated such behavior so far.

Our analysis of Russian foreign policy and military behavior in the Arctic is further informed by a review of Russia's Arctic policy and doctrine; secondary sources on Russia's Arctic Strategy and foreign policy; and conversations, including during a roundtable, with subject-matter experts knowledgeable about Russian Arctic policy. Our analysis also relies on several assumptions: First, we assume that Russia is a rational actor that tries to make optimal choices in the Arctic; i.e., maximizing its benefits while minimizing costs.[1] "Benefits" are defined as those objectives Russia is seeking to achieve in the Arctic based on its official Arctic doctrine. In particular, this chapter examines how systemic factors (for instance, climate) might increase costs for Russia. Second, we recognize that Russia is not a unitary actor, as President Vladimir Putin's diplomatic stance is constrained to some extent by the necessity to satisfy—or at least not antagonize—domestic audiences, a notion that was popularized by Robert Putnam's seminal work on two-level games.[2] However,

[1] For a seminal work on the rational actor model applied to international security issues, see Thomas Schelling, *Strategy of Conflict*, Cambridge, Mass.: Harvard University Press, 1963.

[2] Robert Putnam, "Diplomacy and Domestic Politics: The Logic of Two-Level Games," *International Organization*, Vol. 42, No. 3, Summer 1988.

we also recognize that Russia's Arctic policy is a centralized process in which the Kremlin—and Putin in particular—play an outsized role.[3] Third, we borrow from liberal institutionalism theory when we note that Russia's membership in many Arctic-centered institutions creates conditions for more cooperation by providing forums where potential disagreements may be discussed and solved before they escalate. Still, we make this assumption secondary to the rational actor assumption, noting that Russia could in theory pull out of these institutions if it were to believe its needs were not being met anymore. Fourth, we agree with Russia expert Marlène Laruelle that military and economic concerns are deeply intertwined in the Arctic and that these concerns can, at times, lead to apparently disjointed Russian policies in the region.[4]

This chapter finds that while Russia's current remilitarization of its Arctic region does not necessarily portend an increased risk of conflict, Russia's actions and rhetoric with regard to the Arctic have been inconsistent, alternating in recent years between inflammatory and cooperative. While cooperation has been maintained successfully over the years, none of the factors that made it in Russia's interest to follow international norms and work with Arctic partners will necessarily hold if circumstances change. As a result, Russia's intentions in the Arctic—whether belligerent or peaceful—remain uncertain, particularly after the 2014 Russian intervention in Ukraine that largely came about as a surprise and has led to drastic reassessments of how well we can understand, and predict, Russia's actions throughout its area of influence, which includes a large portion of the Arctic.[5]

---

[3] Marlène Laruelle, *Russia's Arctic Strategies and the Future of the Far North*, New York: M.E. Sharpe, Inc., 2014, pp. 6–9.

[4] Laruelle, 2014, pp. 7–8.

[5] See, for instance, Magnus Christiansson, *Strategic Surprise in the Ukraine Crisis: Agendas, Expectations, and Organizational Dynamics in the EU Eastern Partnership Until the Annexation of Crimea 2014*, thesis, Stockholm, Sweden: Swedish National Defence College, August 2014; Kristin Ven Bruusgaard, "Crimea and Russia's Strategic Overhaul," *Parameters*, Vol. 44, No. 3, Fall 2014, p. 84; and Heather A. Conley, "Russia's Influence on Europe," in Craig C. Cohen and Josiane Gabel, eds., *2015 Global Forecast: Crisis and Opportunity*, Washington, D.C.: Center for Strategic and International Studies, 2014, p. 28.

Russia's Increased Military Activity in the Arctic

Russia is engaged in a large and sustained effort to rebuild and modernize its military capabilities, including in the Arctic, raising questions as to whether this should be seen as merely a legitimate attempt to secure an extensive and increasingly exposed coastline or as a military buildup that announces some aggressive intent. We discuss Russia's Arctic military activity in the context of its overall defense modernization efforts at the end of this section. Here, we first consider Russia's military modernization program in the Arctic through the lens of what Moscow perceives as threats and security priorities.

First, Russia sees maintaining its nuclear deterrence capability as a key strategic priority.[6] This is particularly relevant for the Arctic Ocean, which hosts more than two-thirds of Russia's sea-based nuclear warheads.[7] Historically, the Arctic has been a prime location for Russia's global strategic assets for two reasons: It is the shortest flight path for missiles toward the United States and the Arctic offers good access to the Atlantic.[8] Indeed, the melting of sea ice will soon provide Russia with seasonal strategic links among the Atlantic, Arctic, and Pacific oceans. Russia's aim to maintain its strategic deterrent has been a constant tenet of its security strategy, even in the fiscally difficult years that followed the collapse of the Soviet Union. It remains a priority to this day and a key element to Russia's ambition to be viewed as a great power, suggesting that Moscow will continue to invest in these Arctic-

---

[6] Sophia Dimitrakopoulou and Andrew Liaropoulos, "Russia's National Security Strategy to 2020: A Great Power in the Making?" *Caucasian Review of International Affairs*, Vol. 4, No. 1, Winter 2010, p. 39; Olga Oliker, *Russia's Nuclear Doctrine: What We Know, What We Don't, and What That Means*, Washington, D.C., Center for Strategic and International Studies, May 2016, pp. 2–4.

[7] Data as of 2011. Jørgen Staun, *Russia's Strategy in the Arctic*, Copenhagen, Denmark: Royal Danish Defence College, March 2015, p. 26.

[8] Katarzyna Zysk, "The Evolving Arctic Security Environment: An Assessment," in Stephen Blank, ed., *Russia in the Arctic*, Carlisle, Pa.: U.S. Army War College, July 2011, p. 111; conversation with Danish defense official, Copenhagen, January 2016.

based capabilities used to support its global agenda, regardless of its economic and financial situation.[9]

Second, Russia—like any other state—places a high value on territorial security and seeks to deter, and prepare for, both state and nonstate threats against its strategic infrastructure, whether military or economic.[10] Russia's 2013 Arctic Strategy mentions ensuring national security, protection, and defense of the state border in the Arctic as one of six development and national security priorities in the Arctic zone of the Russian Federation.[11] The Strategy also highlights the need to "ensure a favorable operational regime in the Arctic zone of the Russian Federation, including the maintenance of the necessary combat potential of the general purpose troops of the Russian Federation Armed Forces, as well as other types of troops, military units, and agencies in the region."[12] The diminishing seasonal sea ice that used to protect Russia's northern border has created some concerns for Russia that both states and nonstate actors can now launch attacks against critical infrastructure from surface ships and from underwater in areas that used to be covered in ice.[13]

Finally, Russia has historically been particularly wary of encirclement—a concern amplified by the fall of the Soviet Union, which removed overnight the security brought by a string of buffer

---

[9] Conversations with Arctic experts, Oslo and Copenhagen, January 2016; Zysk, 2011, p. 113; Barbora Padrtová, "Russian Military Build-Up in the Arctic: Strategic Shift in the Balance of Power or Bellicose Rhetoric Only?" *Arctic Yearbook 2014*, Northern Research Forum and the University of the Arctic Thematic Network on Geopolitics and Security, 2014, p. 6.

[10] Katarzyna Zysk, "Russia and the Arctic: 'Territory of Dialogue' and Militarization," briefing presented at the Arctic Frontiers Conference, Tromsø, Norway, January 29, 2016.

[11] The other five priorities are socioeconomic development, science and technology, information infrastructure, environmental safety, and international cooperation. "Стратегия развития Арктической зоны Российской Федерации и обеспечения национальной безопасности на период до 2020 года [Strategy for Development of the Arctic Zone of the Russian Federation and Ensuring National Security for the Period Until 2020]," Moscow, Russia: Russian Federation, February 2013.

[12] "Стратегия развития Арктической зоны Российской Федерации и обеспечения национальной безопасности на период до 2020 года [Strategy for Development of the Arctic Zone of the Russian Federation and Ensuring National Security for the Period Until 2020]," 2013, p. 15.

[13] Zysk, 2011, p. 111; conversation with Arctic experts, Copenhagen, January 2016.

states between Russia and the West.[14] The Arctic, as a long, seasonally ice-covered coastline, protects Russia's north. From its position in the Kola Peninsula in northwestern Russia, the Northern Fleet can access the Atlantic without the proximity of NATO countries—unlike the Baltic and Black Sea Fleets, which have to navigate close to NATO members to reach the Atlantic.[15] This notion of encirclement is also present in the perception by Russia that other Arctic states are leagued against it to limit its presence in the Arctic.[16] Reflecting this concern, Deputy Prime Minister in charge of Defense Dmitry Rogozin noted an emphasis on the Arctic and Atlantic in his report to Putin on the 2015 Maritime Doctrine.[17]

Russia's military programs in the Arctic reflect these three concerns. One priority has been to increase presence by rebuilding a number

---

[14] Olga Oliker, Christopher S. Chivvis, Keith Crane, Olesya Tkacheva, and Scott Boston, *Russian Foreign Policy in Historical and Current Context: A Reassessment*, Santa Monica, Calif.: RAND Corporation, PE-144-A, 2015, p. 5; Staun, 2015, pp. 18–20.

[15] Matthew Bodner, "New Russian Naval Doctrine Enshrines Confrontation With NATO," *Moscow Times*, July 27, 2015b.

[16] See Laruelle, 2014, p. 11; Ekaterina Klimenko, *Russia's Arctic Security Policy: Still Quiet in the High North?* Solna, Sweden: Stockholm International Peace Research Institute, Policy Paper 45, February 2016, p. 35. All Arctic coastal states except Russia are members of NATO.

[17] In the report, Rogozin said,

> Mainly, two directions are emphasized: the Arctic and Atlantic directions. The reasons are as follows. Atlantic is emphasized because lately there has been a rather active development of NATO and its encroachment to Russian borders. Secondly, now that the Crimea and Sebastopol are reunited with the Russian Federation, it is essential to take steps for their integration into the economic activity of the Crimea and Sebastopol. Also, re-establishing the presence of the Russian sea fleet in the Mediterranean Sea. As for the Arctic, the focus on it is explained by several reasons: the growth in significance of the Northern Sea Route, unobstructed access to the Atlantic and the Pacific oceans, and, of course, the wealth of the continental shelf.

"Морская доктрина Российской Федерации: Владимир Путин провёл совещание, на котором обсуждалась новая редакция Морской доктрины Российской Федерации [Maritime Doctrine of the Russian Federation: Vladimir Putin Held a Conference to Discuss the New Edition of the Maritime Doctrine of the Russian Federation]," transcript, Moscow, Russia: Kremlin.ru, July 26, 2015. See also Klimenko, 2016, p. 16; Nikolai Novichkow, "Russia's New Maritime Doctrine," *IHS Jane's Defence Weekly*, August 14, 2015; "Russia Sees Arctic as Naval Priority in New Doctrine," BBC, July 27, 2015.

of Cold War–era bases along with constructing new ones, including on the New Siberian Islands, Wrangel Island, and Cape Schmidt, all along the Northern Sea Route. An Arctic Brigade was created out of the 200th Motor Rifle Brigade in Pechenga and based in Alakurtti, close to the Finnish border; a second one is in the plans.[18] Both brigades should receive navy and air components by 2020.[19] Russia is also investing in new polar-ready equipment, including three new nuclear-powered and four diesel-powered icebreakers, much needed in a fleet that is aging fast under the harsh conditions of the Arctic. Russia is also increasing both its domain awareness and lines of defense. In 2008, surface naval patrols to the Arctic Ocean resumed after an interruption of almost two decades.[20] Russia is building ten air-defense radar stations and announced it would install S-400 air defense missiles on the Novaya Zemlaya archipelago and in the port of Tiksi, and plans to deploy MiG-31 interceptors.[21] At the organizational level, Russia reorganized its military command structure by creating in December 2014 a Northern Joint Strategic Command based in Murmansk to coordinate all military assets in the Arctic region, including the Northern Fleet, which had previously been divided among three different commands.

Russia has also been developing its capabilities for civil response, with a planned increase in the number of search-and-rescue (SAR) stations along the Northern Sea Route and procurement projects for the Russian Coast Guard.[22] These capabilities will also be of use against such threats as terrorism or illegal migration, which Russia plans to

---

[18] Klimenko, 2016, p. 22.

[19] Märta Carlsson and Niklas Granholm, *Russia and the Arctic: Analysis and Discussion of Russian Strategies*, Stockholm, Sweden: Swedish Defence Research Agency, March 2013, p. 26.

[20] Christian Le Mière and Jeffrey Mazo, *Arctic Opening: Insecurity and Opportunity*, London: International Institute for Strategic Studies, 2013, pp. 86–87.

[21] Le Mière and Mazo, 2013, pp. 27–28; Trude Pettersen, "Russia Sends Mig-31 Interceptors to the Arctic," *Barents Observer*, September 25, 2012b.

[22] Padrtová, 2014, p. 5; Katarzyna Zysk and David Titley, "Signals, Noise, and Swans in Today's Arctic," *SAIS Review of International Affairs*, Vol. 35, No. 1, Winter–Spring 2015, p. 174; Matthew Bodner, "Russia's Polar Pivot: Moscow Revamps, Re-Opens Former Soviet Bases to Claim Territories," *Defense News*, March 11, 2015a; Tom Parfitt, "Russia Sends Troops and Missiles to Arctic Bases," *The Times* (UK), December 26, 2015; Klimenko, 2016, p. 25.

counter through the creation of Arctic border guard units in Arkhangelsk and Murmansk and a counterterrorism center in Murmansk.[23] In 2013, Russia also established the Northern Sea Route Administration, which will be responsible for Arctic shipping procedures, including matters related to environment and security.[24] This move also has created increased transparency about Northern Sea Route tariffs and regulations, decreasing uncertainty for shippers and other mariners interested in transiting the waterway.

The scale and frequency of Russia's military exercises have increased, both in the Arctic and in other regions, since 2009.[25] Some exercises in the Arctic have clearly focused on protecting infrastructure, such as the June 2014 exercise that simulated responding to a terrorist attack against an oil terminal near the Pechora Sea,[26] and Russian officials have generally emphasized the defensive nature of these exercises.[27] But other

---

[23] Katarzyna Zysk, "Russia's Arctic Strategy: Ambitions and Constraints," *Joint Force Quarterly*, No. 57, second quarter, 2010, p. 107; "Russia Prepares for Arctic Terrorism," *Maritime Executive Newsletter Online*, December 31, 2015. For a detailed description of Russia's existing and planned military capabilities in the Arctic, see Laruelle, 2014, pp. 113–134; Padrtová, 2014; Klimenko, 2016, pp. 31, 36; conversation with Arctic experts, Oslo, January 2016; Staun, 2015, pp. 24–26; Zysk, 2011; and Kristian Atland, "Russia's Armed Forces and the Arctic: All Quiet on the Northern Front?" *Contemporary Security Policy*, Vol. 32, No. 2, August 2011.

[24] See, for instance, Atle Staalesen, "Opening the Northern Sea Route Administration," *Barents Observer*, March 21, 2013.

[25] While Zapad-09 was at the time the largest exercise conducted since 1991 with 12,500 participants, exercise Vostok-2014 involved 155,000 participants by 2014 (Johan Norberg, *Training to Fight: Russia's Major Military Exercises 2011–2014*, Stockholm, Sweden: Swedish Defense Research Agency, December 2015, pp. 11–12). These exercises are designed to improve military preparedness, and as such are consistent with the reform of the armed forces under way since 2008 to eliminate redundant or hollow units and increase the force's overall level of readiness. Such exercises also play a signaling role: The snap drills and large-scale exercises that took place in the Siberian and Far East military districts are also designed to demonstrate to China that Russia is ready to respond to a potential aggression.

[26] Heather A. Conley and Caroline Rohloff, *The New Ice Curtain: Russia's Strategic Reach to the Arctic*, Washington, D.C., Center for Strategic and International Studies, August 2015, p. 73.

[27] See, for instance, Russian Defense Minister Vladimir Korolyov's statement that the August 2015 exercise in the Arctic involving more than 1,000 soldiers and 14 aircraft was strictly for defensive purposes. "Russia Launches Military Drills in the Arctic," Agence France-Presse, August 24, 2015.

exercises appear more threatening for Russia's neighbors, such as the March 2015 drill that included a takeover of Northern Norway and the seizure of Finland's Åland islands, Sweden's Gotland island, and Denmark's Bornholm island, all located in the Baltic Sea.[28]

Nonetheless, Russia is still a long way from reestablishing the level of military capability it had in the Arctic during the Cold War.[29] While there are certainly new capabilities in the Arctic that are in line with Moscow's various statements on the region's strategic importance, existing capabilities are also eroding fast and Russia's military posture in the region remains quite limited overall.[30] To some extent, this is consistent with Russia's most recent Arctic strategies—the 2008 *Foundation of the State Politics of the Russian Federation on the Arctic for 2020 and in the Longer Perspective* and the *2013 Strategy of the Development of the Arctic Zone and the Provision of National Security until 2020*—which emphasize economic and resources issues over defense and security.[31] The ambitions set out in Russia's *State Armaments Program 2020* will most certainly not be fulfilled within the announced time frame because a number of equipment programs, including for icebreakers and submarines, are already largely behind schedule.[32] This situation will not improve in the short term. The economic recession experienced by Russia, as well as

---

[28] Edward Lucas, *The Coming Storm: Baltic Sea Security Report*, Washington, D.C.: Center for European Policy Analysis, June 2015, p. 9. Some exercises included clear defense and offensive components, such as the 2012 large-scale exercise that allegedly focused on Russia protecting and using the Northern Fleet's strategic submarines to respond to the escalation of a conflict on its southern border. Norberg, 2015, p. 33.

[29] Staun, 2015, p. 26.

[30] Carlsson and Granholm, 2013, pp. 29, 32; Le Mière and Mazo, 2013, p. 87.

[31] Carlsson and Granholm, 2013, p. 15; Alexander Pelyasov, "Russian Strategy of the Development of the Arctic Zone and the Provision of National Security Until 2020 (Adopted by the President of the Russian Federation on February 8, 2013, No. Pr-232)," *2013 Arctic Yearbook*, Northern Research Forum and the University of the Arctic Thematic Network on Geopolitics and Security, 2013.

[32] Atle Staalesen, "Crisis-Ridden Government Cuts Money for Icebreakers," *Barents Observer*, March 16, 2016; Pavel K. Baev, "Russia's Arctic Ambitions and Anxieties," *Current History*, October 2013, p. 266.

recurrent structural issues related to the defense industry,[33] make it likely that many equipment projects will be further delayed.

It is also important to put the new capabilities being developed in the Arctic into a broader context. Russia has engaged in an ambitious, statewide program to restructure and modernize its military and ensure better readiness.[34] Thus, many military developments in the Arctic are actually consistent with changes to the overall Russian defense posture, without signaling any particularly ominous intent in the Arctic. For instance, Russia resumed long-range aviation patrols over the Arctic Ocean in 2007, but also over the Atlantic and Pacific oceans at the same time.[35] It is also worth noting that in spite of the strategic importance of the Northern Fleet, most new naval assets being built or acquired by Russia are destined for other fleets.[36] The Black Sea Fleet, in particular, was given the lion's share of re-equipment projects—a priority that has only been reinforced by the Ukraine crisis.[37] Several analysts have emphasized the difference between militarizing *the* Arctic (to prepare for Arctic-specific threats) and militarizing *in the* Arctic, as part of a larger process of modernization that is more relevant to Russian armed forces in general than to the Arctic in particular.[38] Keeping this dis-

---

[33] Since the Ukraine crisis, Russia has also lost access to the Ukrainian shipyards it previously used for naval construction. See, for instance, Richard Weitz, "Russia's Defense Industry: Breakthrough or Breakdown?" International Relations and Security Network, March 6, 2015.

[34] On Russia's military modernization efforts since 2008, see, for instance, Keir Giles and Andrew Monaghan, *Russian Military Transformation—Goal in Sight?* Carlisle Pa.: Strategic Studies Institute, May 2014; Jim Nichol, *Russian Military Reform and Defense Policy*, Washington, D.C.: Congressional Research Service, R42006, August 24, 2011; and Gustav Gressel, *Russia's Quiet Military Revolution, and What it Means for Europe*, London: European Council on Foreign Relations, policy brief (ECFR/143), October 12, 2015.

[35] Le Mière and Mazo, 2013, p. 86.

[36] Le Mière and Mazo noted that "This shift in focus away from the Northern Fleet towards the smaller organizations reflects a change in Russia's strategic posture as it prioritises its Asian commitments, tries to secure its southern borders and sees NATO as less of a threat." Le Mière and Mazo, 2013, p. 85.

[37] Carlsson and Granholm, 2013, p. 28; Klimenko, 2016, p. 28.

[38] Klimenko, 2016, pp. 31, 34, and 36; Zysk, 2011, p. 112; conversation with Danish defense official, Copenhagen, January 2016; conversation with Arctic experts, Oslo, January 2016.

tinction in mind is important insofar as it puts increased military activity in the Arctic in a broader perspective and suggests that the Arctic may not warrant more concern than other regions of Russia that are also receiving more attention, in terms of military assets and activities, from the Kremlin.

Even so, it is worth considering whether new Russian military capabilities in the Arctic could be used for a potential conflict in the Baltic Sea. Overall, it seems likely that Russia's new military capabilities in the Arctic would play a secondary role in a Baltic contingency.[39] While Russia's air-mobile forces and air forces based in the Arctic theoretically could be sent anywhere, the ground forces component in the Arctic is relatively small and geared toward operations in that region, so the effort required to transport these units does not seem worth the trouble when others are available to support Baltic operations (BALTOPS). The Murmansk-based 200th Independent Motor Rifle Brigade was deployed previously in Ukraine, so the use of parts of that Brigade in a Baltic contingency is plausible, yet unlikely to make a clear difference to the fight, also considering the inherent time and cost in deploying to the Baltic region. The Northern Fleet would have an important role in a Baltic contingency—its forces could defend Russia's Arctic-based nuclear forces, and could menace NATO surface vessels in the North Atlantic. These naval assets could cause some disruption to NATO operations but likely would not contribute much combat power directly to operations in the Baltic states. The missiles deployed in the Arctic are mainly long range and nuclear armed; they would be of little to no use in a conventional Baltic scenario. Those air defenses in the Arctic, including S-300 and S-400 long-range surface-to-air missiles, which would in turn be defended by point defenses like the Pantsir-S1, probably would be kept in their Arctic locations so as not to undermine Russia's ability to defend its nuclear assets.

Another scenario worth considering is whether Russia's enhanced Arctic capabilities might be employed should tensions in the Baltic

[39] Conversation with Arctic experts, Copenhagen, January 2016; conversation with Ministry of Defense official, Copenhagen, January 2016; conversation with Scott Boston, RAND Corporation, July 2016.

region spill over into the Arctic. Unlike Ukraine, another area of high tension with Russia, the Baltic region is geographically close to the Arctic and some of the other Arctic states are geopolitical players in the region, which makes this situation plausible, if not likely. While the Baltic Sea area has seen intense military-related activity since 2014—including a much higher number of Russian air patrols in the area, as well as suspected Russian submarines getting close to Helsinki and the Swedish coast—this is much less true of the Arctic. Russian jet overflights of Norway have increased, but ten times less than similar overflights in the Baltics.[40] The numbers of such overflights have also fluctuated through the years, regardless of activity in the Baltic, increasing sharply between 2006 and 2008, decreasing afterward, and picking up again after 2012.[41] Amid increased tensions with NATO in the Baltic region, Russia did conduct military exercises in the Arctic.[42] NATO followed with its own Arctic military exercise, led by Norway. Such exercises may not present an immediate threat to cooperation in the region, but they could continue to heighten tensions and potentially lead to misunderstandings that result in conflict escalation.

Russia's Rhetoric on the Arctic: Navigating Between Extremes

Russia's discourse on the Arctic has mostly taken a cooperative tone. Russian authorities have also repeatedly argued against a competitive, militaristic vision of the Arctic,[43] and Rowe and Blakkisrud's analysis

---

[40] John Rahbek-Clemmensen, "Carving up the Arctic: The Continental Shelf Process Between International Law and Geopolitics," *Arctic Yearbook 2015*, Northern Research Forum and the University of the Arctic Thematic Network on Geopolitics and Security, 2015, p. 336.

[41] Alexander Sergunin and Valery Konyshev, "Russian Military Activities in the Arctic: Myth and Realities," *Arctic Yearbook 2015*, Northern Research Forum and the University of the Arctic Thematic Network on Geopolitics and Security, 2015, p. 405; Rahbek-Clemmensen, 2015, p. 336; Le Mière and Mazo, 2013, p. 86.

[42] Damien Sharkov, "NATO and Russia 'Preparing for Conflict,' Warns Report," *Newsweek*, August 12, 2015.

[43] Zysk and Titley, 2015, p. 174; Klimenko, 2016, p. 2.

of media coverage of the Arctic by the Russian government–owned newspaper *Rossiiskaya gazeta* from May 2008 to June 2011 shows a similar dominance of cooperative over competitive tone.[44] Both 2008 and 2013 Russian Arctic strategies mention at length the importance of keeping the Arctic as a zone of peace and cooperation.[45]

Yet some high-profile events and statements that present a much more assertive and provocative tone have garnered media attention. Two particularly noteworthy events were the planting of a Russian flag on the bottom of the sea at the North Pole by Artur Chilingarov (a polar researcher and special representative of President Putin for the Arctic) and the unplanned visit to the Svalbard Archipelago of Rogozin, deputy prime minister and head of Russia's Arctic Commission, in April 2015.[46] Chilingarov stated about the Arctic: "Historically speaking, it is Russian territorial waters and islands. Now, we are recovering it."[47] Rogozin declared the Arctic to be "Russia's Mecca"[48]—remarks that are reminiscent of rhetoric used by the Russian power to justify its aggression against Crimea.

One possible explanation for this duality of Russia's official discourse is that the most inflammatory statements are likely aimed at the Russian domestic audience, particularly the nationalistic-inclined

---

[44] Elana Wilson Rowe and Helge Blakkisrud, "A New Kind of Arctic Power? Russia's Policy Discourses and Diplomatic Practices in the Circumpolar North," *Geopolitics*, Vol. 19, No. 1, 2014, p. 73. These authors also note that "Russian strategy documents relating to the Arctic mention a positive 'image' as an important aim for Russia in the Arctic."

[45] "Основы государственной политики Российской Федерации в Арктике на период до 2020 года и дальнейшую перспективу [The Fundamentals of the Russian Federation State Policy in the Arctic for the Period Until 2020 and Beyond]," Moscow, Russia: Government of the Russian Federation, September 18, 2008; "Стратегия развития Арктической зоны Российской Федерации и обеспечения национальной безопасности на период до 2020 года [Strategy for Development of the Arctic Zone of the Russian Federation and Ensuring National Security for the Period Until 2020]," 2013.

[46] Svalbard is under Norwegian sovereignty but ruled by the 1920 Spitsbergen Treaty ratified by 42 countries.

[47] Artur Chilingarov quoted in Staun, 2015, p. 8.

[48] Ishaan Tharoor, "The Arctic is Russia's Mecca, Says Top Moscow Official," *Washington Post*, April 20, 2015.

part of the Russian electorate that supports Putin.[49] This is not to say that they do not serve a purpose with international audiences as well. As Zysk and Titley note, Russia's more confrontational stance plays a deterrence role and keeps potential geopolitical adversaries at bay.[50] The importance of "strategic deterrence in the times of peace" is actually mentioned in Russia's 2013 Arctic Strategy.[51] More generally, such statements are also a way to remind the world that Russia is a great power—in the Arctic and elsewhere—and should be regarded as such.

A second potential explanation for this duality is that the most inflammatory statements are used by individuals to further their own political ambitions (later endorsed by the Kremlin[52]), rather than representing Russia's long-term Arctic policy.[53] Chilingarov's expedition that led to the flag-planting stunt was privately funded, and Byers notes that he was also at the time, "a member of the Russian Duma, in the midst of an election campaign."[54] Additionally, Foreign Minister

---

[49] Laruelle, 2014, p. 3; Dmitry Gorenburg, "How to Understand Russia's Arctic Strategy," *Washington Post*, February 12, 2014a; conversation with Arctic expert, Oslo, January 2016; Padrtová, 2014, p. 1.

[50] Zysk and Titley, 2015, p. 174.

[51] Specifically, this strategy calls for a

> comprehensive military and mobilization readiness at the levels necessary to prevent any military pressure and aggression against the Russian Federation and its allies, to ensure Russian sovereignty in the Arctic, and possibilities for unobstructed activities in all spheres, including in the exclusive economic zone and on the continental shelf of Russia in the Arctic, elimination of the internal and external military threats and providing strategic deterrence in the times of peace, and cessation of military activities in accordance with Russia's interests in times of war.

"Стратегия развития Арктической зоны Российской Федерации и обеспечения национальной безопасности на период до 2020 года[Strategy for Development of the Arctic Zone of the Russian Federation and Ensuring National Security for the Period Until 2020]," 2013, p. 15.

[52] Putin named Chilingarov a "Hero of the Russian Federation," along with two other Russian members of the expedition.

[53] Charles Emmerson, *The Future History of the Arctic*, New York: Public Affairs, 2010, p. 83; Staun, 2015, p. 8; conversation with Arctic expert, Oslo, January 2016.

[54] Michael Byers, *Who Owns the Arctic? Understanding Sovereignty Disputes in the North*, Madeira Park (BC), Canada: Douglas and McIntyre Ltd., 2009.

Sergei Lavrov minimized the incident by comparing it with the U.S. moon landing in 1969—in other words, it was more of a scientific and technological feat than a means to assert sovereignty, which Lavrov stated would be proved in due time, with scientific evidence, before the CLCS.[55] Baev goes even further by suggesting that Putin's interest in the Arctic followed, rather than prompted, the Chilingarov stunt.[56] While the Kremlin did not disavow these statements and events, the fact that it did not follow up on them with more inflammatory rhetoric or subsequent actions—for instance, the "flag on the seabed" event was still followed by a duly submitted claim to the CLCS—suggests that there is indeed a gap between the actions of these individual actors and Russia's long-term policy.

Finally, it could simply be that Russia engages in discourse and rhetoric that best serves its purposes at the time. Sending what seem to be mixed messages is not unusual for Russia or many other world powers, especially when there are many intertwined issues and individuals operating within and outside of the government. Importantly, Russia has thus far appeared to balance or check its aggressive tone to the extent needed to preserve the benefits of long-term regional cooperation, an issue we discuss further in the next section.

Regional Cooperation Has Proven Resilient So Far

In spite of its military buildup and mixed rhetoric, Russia has maintained a cooperative stance overall with other Arctic states over the years. There are numerous examples of cooperation among Arctic nations in the form of political and economic bodies, treaties and agreements, and exercises (See Table 2.1). The sheer diversity of these forums and events—from SAR operations to science and the environment—shows

---

[55] Byers, 2009, p. 88; Mike Eckel, "Russia Defends North Pole Flag-Planting," Associated Press, August 8, 2007.

[56] This author notes that "[Putin] discovered the potential value of the Arctic through the unexpected international resonance from Chilingarov's flag-planting escapade. He was quick to follow up with an order to resume regular long-range aviation patrols over the North Atlantic and North Pacific 'corridors.'" Baev, 2013, p. 269.

Table 2.1
Arctic Nations Have a Strong History of Partnerships and Collaborations

Cooperation Examples[a]	United States	Russia	Canada	Denmark	Iceland	Norway	Sweden	Finland
Arctic Council	√	√	√	√	√	√	√	√
International Maritime Organization	√	√	√	√	√	√	√	√
International Arctic Science Committee	√	√	√	√	√	√	√	√
Arctic Security Forces Roundtable	√	√[b]	√	√	√	√	√	√
NATO	√		√	√	√	√		
Treaties on environment[c]	√	√	√	√	√	√	√	√
Arctic Council SAR Agreement	√	√	√	√	√	√	√	√
UN Convention on the Law of the Sea (UNCLOS)		√	√	√	√	√	√	√
Exercise Cold Response (2006, 2008, 2010, 2012, 2014)	√		√	√		√	√	√
Operation NANOOK (2007–2015)	√		√	√				
SAREX Greenland Sea (2012, 2013)		√	√	√	√	√		

NOTE: Check marks indicate a country's participation.

[a] Collaborative bodies, treaties, and events.

[b] Russia not included in 2014 or 2015.

[c] Includes the Agreement on the Conservation of Polar Bears and the Agreement on Cooperation on Marine Oil Pollution, Preparedness and Response in the Arctic.

the number of areas in which Russia has found it beneficial to work with other Arctic nations. The United States and Russia thus cooperate in a number of activities and agreements that have provided a foundation for generally peaceful bilateral and multilateral relationships on Arctic affairs. The presence of such diverse structures for collaboration creates a kind of diplomatic safety net in the Arctic, which may serve to deter or quell the potential for conflict. Should tensions arise, these collaborative structures may even provide outlets to voice discontent without resorting to overtly aggressive means.

Consensus among Arctic nations has been found even on issues that were initially divisive, such as the decision on whether to grant observer status to non-Arctic nations.[57] Arctic cooperation has also been maintained through difficult times, including the Ukraine crisis that began in 2014. Russia has continued to participate in the work of the Arctic Council, which successfully established in October 2015 an Arctic Coast Guard Forum that will provide all Arctic states, including Russia, with the ability to take part in joint exercises and operations in such areas as SAR and emergency preparedness. Arctic states have also largely maintained their bilateral cooperation with Russia, at least on nonmilitary matters. In March 2015, Norway and Russia carried out their annual "Barents 2015" joint exercise simulating SAR to a vessel in distress and response to an oil spill.[58] The United States still works with Russia on such issues as limitation of black carbon emissions.[59]

The ability to maintain cooperation in the Arctic in spite of a tense international environment dates back to the Cold War. Cooperation even succeeded in areas that would appear to be zero-sum games,

---

[57] While the United States and Norway were in favor of this move, Canada and Russia were wary of having non-Arctic nations influence policy on Arctic issues. At the 2013 Kiruna (Sweden) ministerial meeting, all Arctic Council members agreed to grant Observer status to six more states (China, Italy, Japan, South Korea, Singapore, and India) in addition to the already existing six Observer states (France, Germany, the Netherlands, Poland, Spain, and the United Kingdom).

[58] Trude Pettersen, "Norway and Russia Join Forces in Arctic Response Drill," *Barents Observer*, March 10, 2015.

[59] Victoria Herrmann, "U.S.-Russian Cooperation in the Arctic," *New York Times*, letter to the editor, May 9, 2016.

such as exploitation of resources. One particularly remarkable example was the ability of Norway and the Soviet Union to successfully operate a fisheries management framework in the Barents Sea in spite of Cold War tensions and an (at the time) unresolved territorial dispute. The United States and the Soviet Union discussed and signed the Agreement on the Conservation of Polar Bears in 1973, and Mikhail Gorbachev launched the "Murmansk Initiatives" in 1987 to increase Arctic cooperation on a range of issues from arms control to the environment.[60] The Georgia crisis in 2008 does not appear to have adversely affected cooperation between Russia and other Arctic states.[61]

Cooperation between Russia and its Arctic neighbors, from the Cold War through the development of the Ukraine crisis, has been made possible through a number of factors. Issues affecting the Arctic have for many years been distanced from global politics because it is a remote, ice-covered region with limited short-term prospects for economic gain (although the Soviet Union did develop parts of this region), despite having symbolic and strategic military significance. Put plainly, cooperation has been cheaper than conflict in this region where so much would have to be expended for so little gain.

Cooperation has been further encouraged by the difficulties that Arctic states face governing and operating in such a vast, rigorous, and unforgiving environment. Cold weather and icy conditions combined with limited mobility, communications, and infrastructure make it inherently beneficial to collaborate and make the best of what limited resources are available to cover huge expanses of land and sea. For instance, pooling resources and information for SAR ensures that all nations benefit from life-saving capabilities. The geography and circulation patterns of the Arctic Ocean also quickly make transnational

[60] Lotta Numminen, "Breaking the Ice: Can Environmental and Scientific Cooperation Be the Way Forward in the Arctic?" *Political Geography*, Vol. 29, No. 2, 2010, p. 86.

[61] Elana Wilson Rowe and Helge Blakkisrud, *Great Power, Arctic Power: Russia's Engagement in the High North*, Policy Brief, Norwegian Institute of International Affairs (NUPI), February 2012, , pp. 2–3. If anything, these authors note a higher level of engagement in the Arctic around that time. The effects of the Georgia crisis on international cooperation were limited overall—not just on Arctic issues—as evidenced by the "reset" policy initiated by the United States the following year in its relationship with Russia.

issues out of certain types of threats (such as oil spills or the spread of invasive species from increased maritime traffic), requiring a cooperative response to address them effectively.[62] Collective interests of Arctic indigenous peoples are another set of regional issues that cross national boundaries and foster efforts for collaborations between Arctic nations.

Arctic nations have also largely eschewed attempts to cooperate on potentially fractious issues, such as that of defense. For example, the Arctic Council has been a particularly important institution for maintaining cooperation in times of international crisis. This has been facilitated by the fact that it does not discuss or address military issues. In this forum, Arctic cooperation by its 1996 Charter has focused on environmental protection and sustainable development issues; economic and business interests were moved outside the Council with the creation of the Arctic Economic Forum in 2013. There is no equivalent for military issues, aside from the annual Arctic Chiefs of Defense Staff meetings established in 2012 and suspended in 2014 following the Ukraine crisis, and the Arctic Security Forces Roundtable, which Russia did not attend in 2014 or 2015.[63] Keeping military matters at bay represents both an advantage for the Council, which could continue to operate even as bilateral military relationships between Russia and other Arctic nations were being suspended, and a limitation: Given the lack of civil infrastructure and capability in the Arctic, any major disaster in the region would likely bring military resources together to support civil authorities. While academics and practitioners have been debating on the opportunity of creating such a forum opened to, or focused on, military issues, there has been little interest on the part of Arctic nations to incorporate defense issues into the areas covered by the Council, nor would it necessarily be appropriate, given the pres-

---

[62] Recent years have seen the adoption of a number of international conventions—by the Arctic states or the International Maritime Organization—to regulate such transnational issues for the benefit of all, such as the Agreement on Cooperation on Aeronautical and Maritime Search and Rescue in the Arctic (2011), the Agreement on Cooperation on Marine Oil Pollution Preparedness and Response in the Arctic (2013), or the 2014 International Code for Ships Operating in Polar Waters ("Polar Code").

[63] Klimenko, 2016, p. 30; Dodds, 2016, p. 162.

ence of participating governmental, nongovernmental, and indigenous organizations.[64]

Russia also benefits from stability in the Arctic because it helps enable the economic development and investment it aspires to bring to the region. The difficulty of extracting resources requires large and long-term investments, as well as public-private partnerships to bring the appropriate technology to the areas to be exploited.[65] This last point is of particular importance to Russia, which emphasized in its 2008 Arctic Strategy the "use of Russia's Arctic zone as a strategic resource base of the Russian Federation" to spur socioeconomic development, one of Russia's main national interests.[66] Without cooperation and a generally peaceful environment in the region, Russia would be hard pressed to develop, sell, and transport Arctic natural resources to global markets. High tensions in a region in which it is otherwise so difficult to operate would severely delay or altogether inhibit economic prospects. Russia's emphasis on economic development over defense in the Arctic makes particular sense in the absence of an immediate external threat to economic resources, strategic assets, or influence. These circumstances are vastly different from Ukraine, where Russia perceived a credible threat to its influence and geostrategic interests that seems to have outweighed the economic repercussions.

---

[64] Le Mière and Mazo, 2013, p. 155; Duncan Depledge, "Hard Security Developments," in Juha Jokela, ed., *Arctic Security Matters*, Paris, France: European Union Institute for Security Studies, Report No. 24, June 2015, p. 64.

[65] Baev, 2013, p. 266; Timo Koivurova, Juha Käpylä, and Harri Mikola, *Continental Shelf Claims in the Arctic: Will Legal Procedure Survive the Growing Uncertainty?* Helsinki, Finland: Finnish Institute of International Affairs, Briefing Paper 178, August 2015, p. 5.

[66] "Основы государственной политики Российской Федерации в Арктике на период до 2020 года и дальнейшую перспективу [The Fundamentals of the Russian Federation State Policy in the Arctic for the Period Until 2020 and Beyond]," 2008, p. 2. The 2013 Arctic Strategy outlines steps necessary to secure this particular national interest. "Стратегия развития Арктической зоны Российской Федерации и обеспечения национальной безопасности на период до 2020 года[Strategy for Development of the Arctic Zone of the Russian Federation and Ensuring National Security for the Period Until 2020]," 2013.

CHAPTER THREE

An Examination of Upcoming Transformations in the Arctic

While the previous chapter shows that Russia's track record with regard to Arctic cooperation has been positive, this does not imply that cooperation will necessarily be Putin's preferred option going forward, particularly as the region is experiencing, or could experience within the next few decades, major transformations that could alter Russia's current cooperative stance in the Arctic region and usher in a period of greater tension. Such major transformations include, in particular, maritime access, resources, continental shelf claims, and Russian reaction to NATO presence (Table 3.1).[1]

Table 3.1
Trends Likely to Redefine the Conditions of Cooperation with Russia

Type	Factor	Potential Scenarios
Climate	Arctic sea ice reduction increasing maritime access	Increased access to the Northern Sea Route, prompting concerns from Russia
Economic	Increased global interest in exploiting Arctic resources, and technological ability to do so	Increased international competition to exploit and market Arctic resources
Legal	CLCS decision on continental shelf claims	Russia's claim being denied by UNCLOS Russia using successful claim to overreach
Political/ military	Russia perceives an immediate military threat from NATO in the Arctic	Russia responds with an aggressive move

[1] See Chapter One for how we selected these four factors.

In this chapter, we examine how each of these factors are already under way; how they might further unfold; how they will likely affect Russia's cost-benefit analysis;[2] and what consequences this might have on Arctic cooperation—considering more than one potential scenario when relevant. Although we examine each of these factors individually, the circumstances surrounding them will play out together, almost certainly to an effect that is greater or less than the potential impact of each on its own. For instance, further accelerated melting of sea ice combined with increased global interest in Arctic resources due to market and technological changes could spur Russian aggression in response to a negative (for Russia) CLCS decision. The opposite set of climatic, economic, and technological circumstances could, in turn, minimize Russian reaction to the same CLCS outcome.

Increased Maritime Access Due to Geographical Changes

The highly publicized melting of Arctic sea ice is one fundamental motivation to discuss the region's future, which is why we present it as our first factor. Impacts of a changing climate are being felt in the Arctic sooner and more extensively than almost anywhere else in the world. What was once a place where few adventurers dared to go suddenly seems very nearly within reach. To greater or lesser extents, the potential impacts of all other factors are predicated on assumptions about improved access to the region. Diminishing sea ice is a major enabler for this increased access.[3] Although increases in seasonal maritime access follow a trend that is no longer surprising because these changes have been under way for several decades, it is important to

---

[2] Our analysis of Russia's likely reaction to these factors is based on the same key assumptions discussed in Chapter Two.

[3] Lands in the Arctic are also undergoing incredible physical transformations. However, we do not investigate the very important topics of permafrost melt and coastal erosion. These, too, have significant ramifications for Arctic communities, land-based infrastructure, and access, and should be examined in greater detail. In this report, we are primarily focused on issues at an international scale, which is, perhaps most—but not exclusively—driven by maritime access.

note that this is nonetheless a key transformation because the rate of change and its implications are still being absorbed and understood, and because they have the potential to alter the regional geopolitical landscape—or not, depending on the outcomes of other transformative processes, such as the other factors discussed in this chapter.

Before we can consider the implications of this factor, it is important to summarize what we know about the geography of changing access. This is the subject of numerous academic articles and other reports.[4] Thus, we focus our discussion on some new analysis we have conducted to build upon previous work that centers very directly on questions of maritime access with relevance to Russian plans and intentions in the region.

To aid our analysis, we used a previously developed geographic information system (GIS)–based model called the Arctic Transit Accessibility Model (ATAM), which estimates surface maritime accessibility based on projected sea ice distribution and thickness, as well as assumptions about vessel ice class.[5] We also used several climate projections to estimate changes in access for the next

---

[4] We refer the interested reader to, for example: Julienne Stroeve, Mark Serreze, Sheldon Drobot, Shari Gearheard, Marika Holland, James Maslanik, Walter Meier, and Theodore Scambos, "Arctic Sea Ice Extent Plummets in 2007," *EOS*, Vol. 89, No. 2, January 8, 2008; Yevgeny Aksenov, Ekaterina E. Popova, Andrew Yool, A. J. George Nurser, Timothy D. Williams, Laurent Bertino, and Jon Bergh, "On the Future Navigability of Arctic Sea Routes: High-Resolution Projections of the Arctic Ocean and Sea Ice," *Marine Policy*, Vol. 75, January 2017; Laurence C. Smith, and Scott R. Stephenson, "New Trans-Arctic Shipping Routes Navigable by Midcentury," *Proceedings of the National Academy of Sciences of the United States of America*, Vol. 110, No. 13, March 26, 2013; Charles K. Ebinger, and Evie Zambetakis, "The Geopolitics of Arctic Melt," *International Affairs*, Vol. 85, No. 6, November 2009.

[5] The International Association of Classification Societies has defined different classes of vessels based on their ability to operate in ice-covered water, from Polar Class 1 down to Polar Class 7. See, for example, International Association of Classification Societies, *Requirements Concerning Polar Class*, London, undated.

few decades.[6] The ATAM model itself is described in several peer-reviewed publications.[7]

With the aid of ATAM, we conducted the following analyses:

1. estimating which areas of the maritime Arctic are accessible to vessels with varying physical abilities to operate in sea ice
2. projecting how the open water season for different Arctic coastal locations changes over time
3. examining the impact that Russian denial of access to certain Arctic waters would have on the time required to make a trans-Arctic voyage.[8]

The purpose of these analyses is to provide a better understanding of how maritime access in the region will likely evolve in the coming decades and how Russia might take advantage of this new configuration for economic and strategic advantage. Here, we discuss the results of analyses 1 and 2. We report the findings from Analysis 3 in a later section focused on the upcoming decision on continental shelf claims.

---

[6] For the purposes of this analysis, we ran the model to generate monthly projections of access over the next few decades. We used seven climate projections (ACCESS1.0, ACCESS1.3, CCSM4, GFDL-CM3, IPSL-CM5A-LR, IPSL-CM5A-MR, and MPI-ESM-MR; selected based on availability from previous work and computing capacity within the time frame of the project), and used the Representative Concentration Pathway 4.5, which represents an assumption of moderate climate warming. Although access can change daily and even hourly, we did not have sea ice projections of the resolution required to model at this fine temporal scale. We also did not examine access past 2030 (except for Analysis 3, in which we compare the situation in 2015 with that in 2040) because the next few decades appear most immediately relevant to U.S. government plans. We used multiple climate projections because each representation of climate phenomena comes with its own set of assumptions, strong points, and shortcomings. Having multiple projections ensures that our analysis acknowledges the uncertainty inherent in climate and sea ice representations of the future.

[7] Scott R. Stephenson, Laurence C. Smith, Lawson W. Brigham, and John A. Agnew, "Projected 21st-Century Changes to Arctic Marine Access," *Climatic Change*, Vol. 118, No. 3, June 2013, pp. 885–901; Smith and Stephenson, 2013; Scott R. Stephenson and Laurence C. Smith, "Influence of Climate Model Variability on Projected Arctic Shipping Futures," *Earth's Future*, Vol. 3, No. 1, 2015, pp. 331–343.

[8] We chose a somewhat longer time frame for this analysis because we wanted to maximize the differences in access levels given the data that were available to us.

Figure 3.1 summarizes our results from Analysis 1. These maps compare areas of 90 days of summer access (on average) for open-water vessels and Polar Class 6 ice-breaker vessels for the current decade (2011–2020) and for 2021–2030.[9] The shading in the figure indicates

Figure 3.1
Seven-Projection Estimate for 90-Day Access for Open-Water and Polar Class 6 Vessels from 2011–2020 and 2021–2030

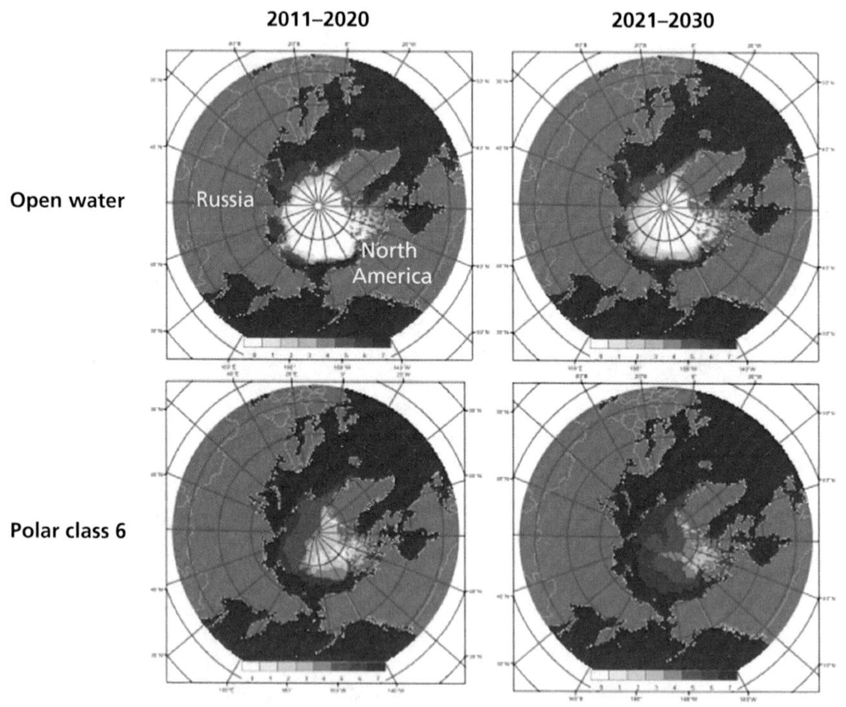

SOURCE: Base map from Central Intelligence Agency, undated.
NOTES: This figure represents summer access spatially averaged at a decadal scale. We calculated the area accessible each year for the greatest 90-day access period between May and December, then averaged it for each decade we looked at. The key at the bottom of each map indicates the number of models that project access to a particular area. White indicates no models project access; dark blue indicates they all do.
RAND *RR1731-3.1*

[9] We did not consider winter access because we (and many others) have found that the Arctic Ocean will remain largely inaccessible during the winter. We chose 90 days somewhat arbitrarily, although it essentially represents a potentially meaningful summer season.

the level of agreement among climate projections that a particular area will be accessible. For example, none of our models predicted continuous access for 90 days or more along the north coast of Greenland, so those areas are white. The nearly continuous dark blue area along the Russian coastline signifies unanimous model projections that large portions of Russia's northern approach will be accessible, even to open-water vessels, by 2030.[10]

It is important to note that future maritime access appears somewhat more permissive during summer seasons, and the Arctic Ocean will remain a seasonally accessible area for all practical purposes. Polar icebreakers may transit the region during wintertime, but these voyages will be largely symbolic or for research. As shown in Figure 3.1, access will increase over time during the warmer months. Some areas, notably around the Canadian archipelago and northern Greenland, appear to remain heavily covered in ice even during the summertime.[11]

The total area of open water (or thinner ice) will increase, as will the length of time that many locations have open water. Figure 3.2 summarizes our results from Analysis 2. Some of these Arctic coastal locations and ports that we examined (Narvik, Murmansk, Nome) are already accessible for the entire 245-day period between May and December.[12] The accessibility of others appears set to increase over time, although there is substantial year-to-year variability that may increase difficulty in planning for access to these locations.[13]

---

[10] Assuming that each climate projection is equally likely (they may or may not be), this is one way of showing confidence in the assessment of increasing seasonal access along Russia's coast. The more models agree, the more confidence there may be in estimates of future access.

[11] We did not explicitly analyze the coverage of first-year versus multiyear ice, although these and other ice characteristics are considered to some extent by the ice multipliers employed within ATAM (see Stephenson and Smith, 2015); we also did not consider daily or hourly patterns of ice cover, which can change rapidly, and are thus very important for navigation.

[12] We define "ports" loosely here; the actual amount of port infrastructure and ship traffic varies widely between these locations.

[13] Note, again, that we did not consider short-term shifts in ice conditions that are very important for navigation. Also, these ports were selected to illustrate variability in the average open-water season, but we might have selected any number of other locations to conduct the same investigation.

Figure 3.2
Number of Days That Selected Arctic Coastal Locations Are Accessible to Open-Water Vessels from May–December 2011–2030

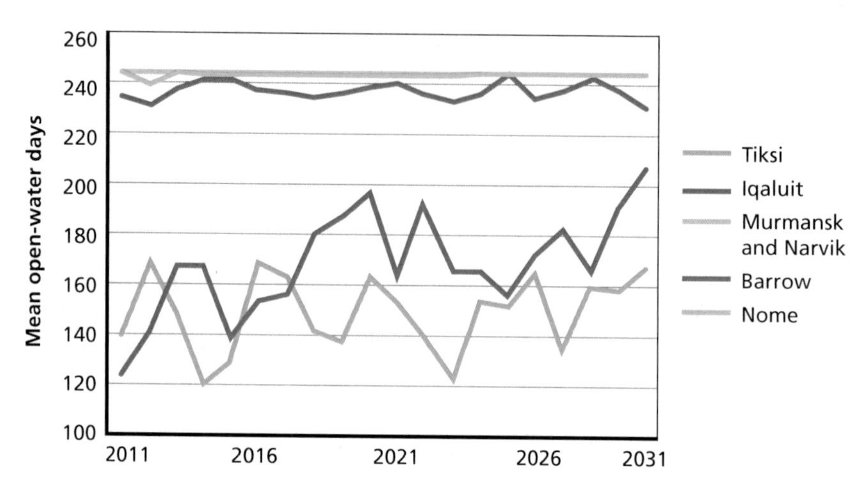

NOTE: Tiksi is in north-central Russia; Iqaluit is in Nunavut, northeastern Canada; Murmansk is in northwestern Russia; Narvik is in Norway; Barrow is in northern Alaska; Nome is in western Alaska.
RAND *RR1731-3.2*

Seasonal access has important implications—especially for Russia, both because of its large Arctic territory and because the increases in access are geographically skewed toward the Northern Sea Route and surrounding areas. Russia can no longer rely quite so heavily on sea ice presenting a natural barrier to foreign operators, and Arctic cooperation could be threatened if continued intense seasonal access changes result in a large uptick in foreign activity along and around the Northern Sea Route. To be clear, it will not be possible for the Northern Sea Route (or any Arctic route) to rival the Suez or Panama Canal without year-round accessibility. However, the potential for some seasonal trans-Arctic traffic, including from and to Asia, is real. Areas along the route will also support resource extraction and transport to global markets. With less seasonal ice to contend with, other government, science, and tourist expeditions may venture closer to Russia's shores.

Not only would foreign presence fuel Russian concerns over sovereignty and encirclement, more activity in general could lead to increased risk for sparking unintended conflicts. The Northern Sea Route is already the focus of disagreement between Russia and other countries (the United States included) as to whether this route can be claimed as Russian internal waters. Without declining seasonal sea ice, debate about transiting the Northern Sea Route without Russia's authorization would continue to be mostly theoretical.[14] A sea lane largely covered in ice has no economic or military value without a sizable icebreaking force—which, arguably, only Russia has.[15] Russia could react with force if the United States and other countries were to take a strong stance on principle to protect freedom of navigation through the Northern Sea Route. However, it is difficult to imagine how a show of force strong enough to elicit a Russian response in kind would serve U.S. national interests in the Arctic, not to mention the fact that such an action would probably require diverting resources used to protect freedom of navigation elsewhere in the world, where stakes are much higher because of more-elevated strategic and economic priorities.

Another plausible risk related to increasing access is that of miscommunication and misinterpretation during routine activities, such as exercises and patrols, which are likely to increase because of the enhanced need for security and monitoring that comes with greater access. The increased interaction of military assets in an area that is not very well serviced by global communications could lead to rapid escalation of individually small incidents.

Increased Interest in Arctic Resources

Increasing Arctic maritime access has raised expectations about unlocking precious energy reserves that to date have been too difficult to extract. In 2008, the USGS raised global interest in the region by

---

[14] Even with the advent of substantially declining seasonal sea ice cover, the route will remain hazardous to navigate because of unpredictable environmental conditions, lack of infrastructure and reliable communications, and limited knowledge about seabed and coastal geography in many areas, among other issues.

[15] Even with a sizable icebreaking force, the route's value would be extremely limited as an economic or military strategic route.

publishing estimates of the yet-untapped reservoirs of oil and natural gas believed to be buried in the Arctic.[16] Could competition for Arctic resources, which not only include hydrocarbons but also minerals and fish, drive the region toward conflict? Russia, in particular, has a high stake in developing Arctic resources, with its lion's share of territory and high hopes for economic progress driven by the exploitation of resources in its northern territories.

We begin by briefly overviewing the Arctic resources that have brought global attention to the region, starting with hydrocarbons. As global energy demand increases and conventional fossil fuel supplies dwindle, the oil and gas industries have begun to seek new reserves in more-extreme environments, such as deep-ocean drilling and exploration in the Arctic.[17] Until recently, most oil and gas exploration and production in the Arctic has been limited to onshore or near-shore operations.[18] Increasingly, companies are prospecting in areas far off shore that were previously determined to be economically infeasible for exploration. These activities could pay off: The 2008 USGS estimates of "undiscovered" Arctic hydrocarbon reserves total 1,669 trillion cubic feet of natural gas, 44 billion barrels of natural gas liquids, and 90 billion barrels of oil.[19] These numbers represent approximately 30 percent of the world's

---

[16] USGS, 2008.

[17] While some major players in the industry are also diversifying their portfolios by investing in renewable and alternative energy technologies, the majority of investment dollars from these companies remain focused on exploitation of fossil fuel resources.

[18] Exploration drilling in the Arctic has been happening for several decades, though only a handful of isolated projects have been carried out on continental shelves. Onshore exploration drilling dates back to the 1940s on the North Slope of Alaska. Near-shore (within 10 miles) activity started in the Beaufort Sea in the 1980s. Offshore exploratory activity began in the Barents Sea in the 1970s in Norway and in the 1980s in Russia. Andrew Rees and David Sharp, *Drilling in Extreme Environments: Challenges and Implications for the Energy Insurance Industry*, London: Lloyd's, 2011; Emily Stromquist and Robert Johnston, *Opportunities and Challenges for Arctic Oil and Gas Developments*, Washington, D.C.: The Wilson Center, 2014; and Ivan Panichkin, "To Explore and Develop," *Russian International Affairs Council*, November 24, 2015.

[19] In May 2008, USGS was the first organization to provide an estimate on potential oil and gas reserves in all areas north of the Arctic Circle (66.56 latitude north). These first estimates, however, are based on very limited geological information and will likely be revised as new

gas and 13 percent of global oil that have not yet been successfully prospected, but appear to exist, geologically speaking, and may be recoverable without relying on major technological breakthroughs.[20]

Across the Arctic region, natural gas is about three times more abundant than oil. Most of the natural gas is in Russia,[21] while the largest amount of "undiscovered" oil may be in the Alaskan Arctic. Most of these resources are projected to exist offshore, on continental shelves in less than 500 meters of water.[22] Russia has submitted a bid to the CLCS to extend its continental shelf (discussed in the next section); if approved, it will provide the country exclusive exploitation rights to continental shelf reserves that may represent up to 60 percent of Russia's total hydrocarbon resources.[23]

Assuming geological estimates are correct, the economic feasibility of the oil and gas buried deep beneath the Arctic Ocean is dependent on a number of factors—principally global energy prices, but also advances in and availability of the infrastructure and technology required to bring these hydrocarbons to market. Environmental and other regulations on operations are also important considerations. Combined, these challenges and constraints make investing in "undiscovered" hydrocarbon prospecting and extraction quite risky. Although many companies are still involved in a number of Arctic drilling projects, Shell's well-publicized exit in 2015 from the North

---

data become available. For details on the assessment methodology, see Donald L. Gautier, Kenneth J. Bird, Ronald R. Charpentier, Arthur Grantz, David W. Houseknecht, Timothy R. Klett, Thomas E. Moore, Janet K. Pitman, Christopher J. Shenck, John H. Schuenemyer, Kai Sorensen, Marilyn E. Tennyson, Zenon C. Valin, and Craig J. Wandrey, "Assessment of Undiscovered Oil and Gas in the Arctic," *Science*, Vol. 324, No. 5931, 2009, pp. 1175–1179; USGS, 2008.

[20] A notable assumption of this study is that offshore resources are technically recoverable under permanent sea ice and any water depth, two known challenges of resource exploitation in the Arctic. This study did not take into account economic considerations.

[21] The USGS report estimates that more than 70 percent of the undiscovered Arctic natural gas reserves exist in three provinces: West Siberian Basin, the East Barents Basins, and Arctic Alaska. USGS, 2008.

[22] Gautier et al., 2009.

[23] Arte Staalesen, "New Reality for Norwegian Defence," *Barents Observer*, April 30, 2015.

Slope of Alaska suggests that not all energy companies are willing to stomach the inherent uncertainty and extremely high cost that go along with Arctic resource extraction.[24]

Mining of metals and rare-earth minerals in the Arctic has great economic potential.[25] Large deposits are known to exist across the Arctic—in Russia, several new mines are being developed along the Yamal Peninsula and other Arctic coastal areas. Greenland's mineral deposits have also gained attention as ice sheets shrink and changes are made to mining policies and regulations.[26] The Red Dog mine in Arctic Alaska is a world leader in producing zinc,[27] and one of the world's largest diamond mines opened in 2016 in northern Canada.[28]

The Arctic also hosts valuable fisheries, which, as climate warms, may become even larger and more diverse if habitats shift northward as predicted. The Arctic coastal states regulate fishing within 200 nautical miles (nm) of their coastlines as well as inland fishing. Management of these fisheries varies; for example, Russia engages in commercial fishing, whereas the United States has prohibited it.[29] There also may be lucrative fishing now, and especially in a warmer future, in the central Arctic Ocean, outside of the coastal states' exclusive economic zones (EEZs).[30] Although fishing in this area is impractical, if not infeasible, due to sea ice, the Arctic Coastal states have recognized its potential draw in the

---

[24] See, for example, Antonia Juhasz, "Shell is Reeling After Pulling Out of the Arctic," *Newsweek*, October 13, 2015.

[25] Heather A. Conley, David Pumphrey, Terrance M. Toland, and Mihaela David, *Arctic Economics in the 21st Century: The Benefits and Costs of Cold*, Washington, D.C.: Center for Strategic and International Studies, July 2013.

[26] Conley et al., 2013; Bryce Gray, "As Greenland Ramps Up Mining, Who Will Benefit?" *Arctic Deeply*, March 17, 2016.

[27] Northern Alaska Environmental Center, "Red Dog Mine," web page, March 26, 2010.

[28] Kate Kyle, "N.W.T.'s Gahcho Kué Diamond Mine Marks Grand Opening Today," CBC News, September 20, 2016.

[29] North Pacific Fishery Management Council, "Arctic Fishery Management," web page, undated; Steven Lee Myers, "Sea Warming Leads to Ban on Fishing in the Arctic," *New York Times*, July 16, 2015.

[30] Myers, 2015.

future for China and other countries that suffer from depleted fish stocks closer to home, and are taking diplomatic and policy steps to ensure that the high seas of the central Arctic remain closed to commercial fishing, at least until more research can be conducted to inform appropriate management for these fisheries.[31] The resulting declaration, signed in 2015, was also a notable diplomatic achievement for Russia and its Arctic coastal neighbors at a time of otherwise strained relationships in the wake of ongoing tensions over Ukraine.

Although there is no doubt that the Arctic contains vast economic potential for Russia and others, there are some important reasons why this factor, on its own, is unlikely to diminish Russia's cooperation with its Arctic neighbors. Potential for high global energy prices, along with the development of the necessary infrastructure and access to extraction technologies, will be instrumental in determining the magnitude of impact from this factor. In addition, Russia's natural gas, oil, minerals, fish stocks, and other resources are not under any major threat, real or perceived, from inside or outside the Arctic. Other than the upcoming CLCS decision we will discuss next, there are no major territorial disputes between Russia and its Arctic neighbors in which there might be substantial resources at stake. Even if Russia is not successful in its bid for certain contested areas of the Arctic Ocean seabed, it has other hydrocarbon resources within its EEZ that can be developed. For example, only about 20 percent of the Barents Sea and 15 percent of the Kara Sea have been explored; the East Siberian Sea, Laptev Sea, and the portion of the Chukchi Sea off the Russian coast have not been explored at all.[32] In the Barents and Kara seas, approximately 430 million tons of oil and 8.5 trillion cubic meters of natural gas have been discovered on the Russian continental shelf but are yet to be developed.[33]

---

[31] Hannah Hoag, "Nations Negotiate Fishing in Arctic High Seas," *Arctic Deeply*, April 28, 2016; Min Pan and Henry P. Huntington, "A Precautionary Approach to Fisheries in the Central Arctic Ocean: Policy, Science, and China," *Marine Policy*, Vol. 63, January 2016, pp. 153–157.

[32] Panichkin, 2015.

[33] Panichkin, 2015.

It could be argued that tensions around resources may come from non-Arctic nations. China, in particular, has shown an interest in Arctic resources. However, Russia has carefully made China a partner, not an adversary, in Arctic resource extraction. For example, the countries are collaborating in efforts to extract natural gas from the Yamal Peninsula. Further, Russia and China appear poised to collaborate, rather than compete, when it comes to mining because it could give them a strategic advantage in controlling rare-earth metals that the United States and other developed countries desperately need for a large number of industries.[34] The Arctic Coastal states, including Russia, have also involved China in discussions about prohibiting commercial fishing in the Arctic high seas, clearly acknowledging its Arctic interests and attempting to secure its compliance with the regulation in so doing.

Destabilizing the region could also limit Russia's potential to benefit from its Arctic resources, which its national priorities clearly indicate it wishes to do. The difficulty of resource exploration and exploitation in this harsh, remote region hugely complicates the road to profitability,[35] let alone if a conflict were to put at risk personnel, ships, infrastructure, and sea lanes needed to support these activities and bring resources to global markets. Further, Russia has long been dependent on western technology for oil and gas exploitation. If sanctions remain in place and western companies advance offshore drilling technology, Russia's access to these innovations may be delayed or prevented altogether, limiting the nation's ability to economically exploit some of its offshore energy reserves.[36]

---

[34] Michael Gorodiloff, "Will Russia and China Set up a Rare Earth Metals Cartel?" *Russia Direct*, February 10, 2016.

[35] Some of the most significant oil and natural gas resources are geologically complicated to extract, which, in combination with volatile global energy markets, has made additional exploration and energy investment in the Arctic largely unfavorable in the near term.

[36] As discussed previously, the sanctions recently imposed on Russia by the United States, the European Union, and other countries have had a negative effect on Russia's ability to exploit Arctic energy resources, at least in the short term. It is estimated that import replacements for current technologies would happen in 2020–2025 at the earliest. Panichkin, 2015.

However, Russia's priorities could change, depending on its strategic circumstances, which could influence its responses to Arctic events. Consider, for example, a situation in which Russia perceives a threat to its Arctic influence and infrastructure from fishing or research vessels flagged to other Arctic countries operating close to or (illegally) just within the Russian EEZ. If Russia were to interdict and hold these in response, tensions could rise and amplify other disagreements, leading Russia to consider suspending cooperation in the Arctic. Russia has displayed a willingness to react with military force to the incursion from another state in its EEZ, particularly if it is perceived as an immediate threat to military or energy infrastructure. For example, in 2013, Russia arrested the crew of Greenpeace vessel *Arctic Sunrise* that attempted to climb its oil rig and charged them first with piracy, then hooliganism (punishable by 15 and 7 years in prison, respectively), while seizing the ship.[37] Earlier in 2012, Russian Border Guard fired at, disabled, and boarded a Chinese vessel illegally fishing in Russia's EEZ.[38]

Upcoming Recommendations on Continental Shelf Claims

There are few territorial disputes in the Arctic, but one of symbolic (and possibly economic) consequence involves sovereignty over the seabed around the North Pole. Russia, Denmark, and Canada all seek to extend the right to exploit their continental shelves northward, and claims about their respective geological boundaries overlap somewhat in the northernmost part of the world. Coastal states normally have the right to exploit their continental shelf only up to 200 nm (i.e., up to the limits of their EEZ), but UNCLOS (of which all Arctic coastal states except the United States are signatories) provides that such rights can be extended if the

[37] Shaun Walker and Sam Jones, "Arctic 30: Russia Changes Piracy Charges to Hooliganism," *The Guardian* (UK), October 23, 2013.

[38] Michael Martina, "China Condemns Russia for Detaining Fishermen," Reuters, July 19, 2012.

continental shelf naturally extends beyond the 200-nm limit.[39] States must then prove that the area they claim as an extension of their continental shelf is geologically similar to the continental shelf closer to their coast, and therefore part of that continental shelf.

Based on the provisions set forth in UNCLOS, the CLCS considers and issues recommendations on sovereignty over the extended continental shelf area based on the scientific evidence brought by coastal states. If two or more claimants have overlapping claims, it is up to these states to work out a delimitation that is mutually acceptable. States are then committed to the final result of this delimitation process, as UNCLOS states that "The limits of the shelf established by a coastal State on the basis of these recommendations shall be final and binding."[40]

Russia first submitted a claim to extend the 200-nm limit of its continental shelf in 2001 and, upon a request from the CLCS for additional scientific evidence, subsequently revised it in 2015. Russia's resubmission on August 4, 2015, reflected many years of additional data-gathering, and its claim now covers more than 463,000 square miles of sea shelf in the Arctic, including area around the North Pole.

Denmark submitted five separate claims between 2009 and 2012, one of which (north of Greenland) overlaps with Russia's claim; Canada made a partial submission in 2013 that mentioned it would submit a complement on the Arctic at a later date. It is expected that Canada's full submission will overlap with Russia's and Denmark's (see Figure 3.3).[41] The overlap in Denmark's and Russia's claims—and Canada's anticipated claim—appear to be, particularly around the North Pole and Lomonosov Ridge. Norway's 2006 submission did not

---

[39] Within the limits set by UNCLOS:

> The fixed points comprising the line of the outer limits of the continental shelf on the seabed, drawn in accordance with paragraph 4(a)(i) and (ii), either shall not exceed 350 nm from the baselines from which the breadth of the territorial sea is measured or shall not exceed 100 nm from the 2,500 metre isobath, which is a line connecting the depth of 2,500 metres.

UNCLOS, Part VI, Article 76, para. 5, Montego Bay, Jamaica, December 10, 1982.

[40] UNCLOS, Part VI, Article 76, para. 8, 1982.

[41] States have ten years after they ratify UNCLOS to prepare a submission to the CLCS.

Figure 3.3
Overlapping Claims in the Arctic

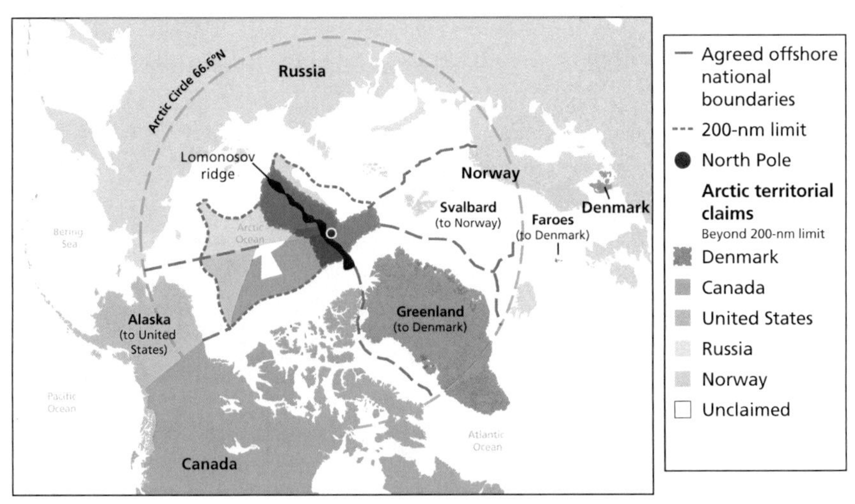

RAND *RR1731-3.3*

overlap with any other claim and received a positive recommendation from the CLCS in 2009.[42] The United States has not submitted any claims because it is not an UNCLOS signatory.

It may still be some time before the CLCS issues any recommendations on these overlapping seabed claims. The commission examines claims from all over the world, and the average time between a submission and a decision is generally three to four years, although the backlog of submissions might stretch this timeline.[43] Further, since there

[42] United Nations, Oceans and Law of the Sea, "Submissions, Through the Secretary-General of the United Nations, to the Commission on the Limits of the Continental Shelf, Pursuant to Article 76, Paragraph 8, of the United Nations Convention on the Law of the Sea of 10 December 1982," United Nations Division for Ocean Affairs and the Law of the Sea, website, October 28, 2016.

[43] Conversation with Arctic expert, Oslo, January 2016. Russia's 2001 claim received a recommendation the following year, while Norway's 2006 claim received a recommendation three years later. It is difficult to say how long it would take for claims submitted at a later date to be adjudicated, but as of January 2017 the Commission had 56 claims that had not yet received a recommendation, 42 of which were still waiting to have a subcommission established for their examination (United Nations, Oceans and Law of the Sea, 2016).

may be three overlapping claims in this case, discussions on the delimitation of the shelf will not begin until the last claim (likely Canada's) receives a recommendation—and then the bilateral or multilateral process for states to agree on a delimitation of their shelves based on the recommendations can take many more years.

The strategic importance of the areas of the Arctic Ocean covered by these claims has been much debated, usually with a focus on oil and gas resources. In that regard, it is important to note that the commission's recommendations only pertain to the rights to exploit the seabed. The column of water above the continental shelf located outside the 200-nm border of coastal states is considered the high seas, which are, according to UNCLOS, "open to all States, whether coastal or landlocked"[44] and administered by the International Seabed Authority. As a result, extended continental shelves have implications for mining and drilling—not navigating or fishing.[45] Hydrocarbons are expected to be present in the areas being claimed, but they are still a long way from being exploitable—and, even then, will still likely be less attractive than the more-accessible and more-abundant reserves that are present closer to shore in undisputed areas.[46] The North Pole, for instance, is not expected to hold large amounts of resources,[47] and even if it did, they would be close to impossible to exploit because they would be sitting at a depth of 4,000 meters.[48]

Even if these areas hold no clear economic or strategic benefit for decades to come, they are still important politically and ideologi-

---

[44] UNCLOS, Article 87, 1982.

[45] Technically, rights to exploit the seabed would extend to the fishing of benthic species (i.e., species that live on the seabed, rather than in the water column above it). Two high-value benthic species that are present in the Arctic are the snow crab and the king crab, but there is too little information available on the quantities and location of these species to speculate about their commercial viability. Some of the contested areas in the Arctic also may be too deep for crab fishing. Conversation with Arctic expert, Oslo, January 2016; Gunnar Knapp, "Arctic Fisheries: Opportunities and Policy Issues," presentation at the UAA Institute of Social and Economic Research, University of Alaska, Anchorage, Alaska, February 27, 2014.

[46] USGS, 2008.

[47] USGS, 2008.

[48] Byers, 2009, p. 93.

cally. The North Pole is a particularly symbolic area for Arctic countries (and, one could argue, for non-Arctic countries as well).[49] While not technically "territories" (states merely gain the right to exploit the seabed, not to claim those parts of the ocean as their own), the areas claimed represent an extension of states' sovereignty. Also, the fact that no one really knows what will be found in these areas 50 or 100 years from now gives states an incentive to push for maximalist claims—within the boundaries of what scientific evidence can support.[50]

These claims also play an important role with regard to domestic politics, particularly for such countries as Russia, Denmark, and Canada that hold the Arctic as an important element of national identity (something we discuss more in the next section). For Denmark, a maximalist claim is a way to show Greenland (whose population has been increasingly supportive of independence from Denmark) that it looks after its interests, by trying to secure the largest possible extension of the continental shelf off Greenland.[51] Canada's 2009 Northern Strategy presents the North as "central to the Canadian national identity" and sets "Exercising Our Arctic Sovereignty" as the first of four objectives outlined in the Strategy.[52]

Although sovereignty disputes have the potential to inflame tensions, thus far Russia, Denmark, and Canada have all dutifully followed the legal process and taken steps to ensure that, even with overlapping claims, they show their commitment to cooperation on continental shelf issues. For instance, Denmark consulted with other Arctic coastal states before submitting its claim in 2014, and exchanged notes with Russia where both states said they did not object to each other's sub-

---

[49] Staun, 2015, p. 10.

[50] Rahbek-Clemmensen, 2015, p. 332.

[51] "Denmark Challenges Russia and Canada Over North Pole," BBC, December 15, 2014; conversation with Arctic experts, Copenhagen, January 2016.

[52] The Strategy defines "Exercising Our Arctic Sovereignty" as "maintaining a strong presence in the North, enhancing our stewardship of the region, defining our domain and advancing our knowledge of the region." Minister of Indian Affairs and Northern Development and Federal Interlocutor for Métis and Non-Status Indians, *Canada's Northern Strategy: Our North, Our Heritage, Our Future*, Ottowa, 2009, p. 9.

mission and would work together on a delimitation line agreeable to both parties.[53] Denmark and Canada also refrained from coordinating their submissions to the CLCS to avoid claiming overlapping areas—an effort on which they had been working for years.[54] Further, Russia has not only respected but also actively supported UNCLOS. When the European Union started promoting an international treaty on the Arctic,[55] Russia and other Arctic coastal states promptly renewed their commitment to UNCLOS through the 2008 Ilulissat Declaration.[56] Russia has also shown some restraint in the claim it submitted, which could have been more extensive that it ended up being.[57] Finally, continental shelf claims have provided further opportunities for Arctic nations to cooperate: Because of the difficulty of gathering the type of scientific data that is required to make such claims, states have worked together through joint expeditions and sharing of equipment to reduce their respective costs and increase their chances of success.[58]

[53] Koivurova, Käpylä, and Mikola, 2015, p. 5. These authors note that this exchange of notes with Russia came after the illegal annexation of Crimea by Russia.

[54] Rahbek-Clemmensen, 2015, p. 335.

[55] The European Parliament suggested in its 2008 Resolution on the Arctic Governance "that the Commission should be prepared to pursue the opening of international negotiations designed to lead to the adoption of an international treaty for the protection of the Arctic, having as its inspiration the Antarctic Treaty." Resolution on Arctic Governance, Brussels, Belgium: European Parliament, October 9, 2008, para. 15. On the arguments for and against an Arctic international treaty, see, for instance, Oran R. Young, "Whither the Arctic? Conflict or Cooperation in the Circumpolar North," *Polar Record*, Vol. 45, No. 232, 2009, pp. 73–82.

[56] Adopted at the Arctic Ocean Conference in Greenland in May 2008, the Ilulissat Declaration, signed by Canada, Denmark, Norway, the Russian Federation, and the United States, states that "Notably, the law of the sea provides for important rights and obligations concerning the delineation of the outer limits of the continental shelf, the protection of the marine environment, including ice-covered areas, freedom of navigation, marine scientific research, and other uses of the sea. We remain committed to this legal framework and to the orderly settlement of any possible overlapping claims." Ilulissat Declaration, *Arctic Ocean Conference*, Ilulissat, Greenland, May 27–29, 2008.

[57] Rahbek-Clemmensen, 2015, p. 336; Baev, 2013, p. 268.

[58] Examples include the Danish-Canadian collaboration on the LOMROG I and LOMROG II continental shelf mapping projects; U.S.-Canada collaboration to map the seabed; a joint paper by Danish and Russian geologists on how the seabed connects with the onshore part of the con-

Despite displaying restraint in its claim, Russia has made clear how much it values the delimitation of its continental shelf in its successive Arctic strategies. While the 2008 Strategy talks about the need to explore and study the continental shelf to increase the reserves of mineral resources, the 2013 Strategy names demarcation of the continental shelf as an immediate goal, specifically in "prevention of spatial loss and preclusion of unfavorable legal conditions for Russia's activity in the Arctic as compared to other Arctic states."[59]

For Russia, the symbolic value and prestige of the continental shelf claim, combined with the increasingly nationalistic tone adopted by the Kremlin, suggest that the CLCS decision could have an important impact if Russia's claim is denied. Denmark and Canada could then start negotiating a delimitation of their respective extended continental shelf, likely including the areas where their claims overlapped with those of Russia.

If its claim were denied, one option for Russia would be to resubmit.[60] This would involve undertaking additional scientific research to find more data, as it did between 2002 and 2015—an extremely costly endeavor, made more difficult by the severe economic recession that Russia is experiencing as of 2016. Politically, Russia may also be disinclined to submit a third claim after having been denied twice, especially as there is no reason to believe that they have not provided all the evi-

---

tinental shelf; the use by Canada of a Russian icebreaker in 2007 along the coast of Greenland for mapping purposes; and the signing in December 2007 of an agreement between Canada and Russia recognizing the need to cooperate to map the Arctic Ocean. Numminen, 2010, p. 86; Byers, 2009, pp. 95–97; conversation with Geological Survey of Denmark and Greenland management, Copenhagen, January 2016.

[59] "Основы государственной политики Российской Федерации в Арктике на период до 2020 года и дальнейшую перспективу [The Fundamentals of the Russian Federation State Policy in the Arctic for the Period Until 2020 and Beyond]," 2008, p. 4; "Стратегия развития Арктической зоны Российской Федерации и обеспечения национальной безопасности на период до 2020 года [Strategy for Development of the Arctic Zone of the Russian Federation and Ensuring National Security for the Period Until 2020]," 2013.

[60] The CLCS typically would not tell a given claimant it has no rights to the continental shelf outlined in its submission. Rather, its recommendations focus on whether the scientific evidence submitted is sufficient to establish such rights. As a result, a state that receives a negative recommendation can choose to resubmit a claim.

dence they could provide this time. Yet this option cannot be entirely ruled out, especially because new technologies, particularly in the field of deep-sea drilling (which could be used to reveal additional geological evidence), might turn up new evidence that would make a better case.[61]

Alternatively, geopolitical tensions could intensify if Russia rejects the commission's recommendation and refuses to resubmit a claim, deciding instead to increase military presence in the region to deter Canada and Denmark from trying to exploit their newly extended continental shelf rights. Russia could also interdict contested areas to Danish and Canadian exploratory teams, betting on the fact that neither Denmark nor Canada is ready to go to war for a section of the Arctic seabed. Backing such moves by military force may present a certain risk because both Denmark and Canada are NATO members. Yet an activation of the collective defense principle in this case seems highly unlikely. The Arctic of minor importance to most NATO members, and the question of whether Article 5 of the Washington Treaty covers an extended continental shelf—which is not part of a state's territory per se—is highly debatable. Russia would probably be correct to assume that the North Atlantic Council (NAC) would have a difficult time coming to a consensus on this issue, which to a large extent negates the deterrent effect that NATO membership would normally have on Russia.

It is also possible that Russia could simultaneously announce plans to resubmit a claim and harass Denmark and Canada as they engage in bilateral discussions in the hopes of slowing their progress and deterring their presence around the North Pole. Russia might even impede negotiations long enough to come up with more scientific data that would support its previously denied claim. The commission has to focus on the scientific evidence for continental shelf extensions; thus, it cannot deny a third Russian submission because of belligerence toward its Arctic neighbors.

---

[61] According to a January 2016 conversation with Geological Survey of Denmark and Greenland management, there is no "stigma" to resubmitting a claim if a state increases the area it claims or if it comes up with additional scientific evidence, as there is a general understanding that the Arctic is a very difficult zone to map and survey for geological data.

Despite the lack of concrete military or legal deterrents to Russia exhibiting discontent with the progress of the CLCS determination, there are also incentives for it to not openly contest an UNCLOS-based decision. Doing so would be risky for Russia, as it might open a "Pandora's box" whereby other decisions—including the 2014 CLCS recommendation backing Russia's claim in the sea of Okhotsk—could be contested by third parties. Creating such uncertainty is not in Russia's interest, particularly for seabed areas that are mostly worthless until market conditions and technological advances make their exploitation economically viable.[62] Russia also benefits from UNCLOS in the Arctic in the sense that the current regulation makes it clear that most of the Arctic seabed can only be claimed by Arctic coastal states. If that rule were to disappear, other states would probably want to enter the competition for exploitation rights[63]—China, for instance, has shown sustained interest in Arctic resources.[64]

The United States also might have grounds for concern if the commission were to decide in Russia's favor. In this case, Russia could choose to overreach and grab not only exploitation rights to the seabed but also attempt to control access on the surface by interdicting or limiting the transit of international vessels. This is not far-fetched, considering Russian policy in the Northern Sea Route. In addition to claiming that some of the route's straits are internal waters, Russia has chosen to interpret UNCLOS Article 234 (which gives coastal states "the right to adopt and enforce non-discriminatory laws and regulations for the prevention, reduction and control of marine pollution from vessels in ice-covered areas within the limits of the exclusive economic zone"[65]) as authorizing it to control transit passage—even as the Northern Sea

---

[62] Conversation with Arctic experts, Oslo, January 2016.

[63] Conversation with Arctic expert, Oslo, January 2016.

[64] See, for instance, Erik Solli, Elana Wilson Rowe, and Wrenn Yennie Lindgren, "Coming into the Cold: Asia's Arctic Interests," *Polar Geography*, Vol. 36, No. 4, 2013; Tim Boersma and Kevin Foley, *The Greenland Gold Rush: Promises and Pitfalls of Greenland's Energy and Mineral Resources*, Washington, D.C.: Brookings Institution, September 2014.

[65] UNCLOS, Article 234, 1982.

Route becomes less and less ice-covered.[66] Such an overreach by Russia over the North Pole would not only be destabilizing for Arctic security, but might also be used as a precedent by states in other regions to assert their own claims—for instance, China in the South China Sea.[67]

In the event that Russia decides to overreach and control maritime activity around the North Pole, it could interfere with the use of a polar maritime route that currently has little but symbolic significance, but could develop strategic, military, and economic importance as seasonal access continues to increase.[68] The impacts of any future denial of access vary, according to our ATAM analysis, depending on the icebreaking capability of the vessel in question.[69] Although a U.S. icebreaker in 2040 could navigate around Russia's EEZ and outer continental shelf with no appreciable impact on the length of its trans-Arctic route, vessels with no special icebreaking capability could be forced to take routes that are between 45 percent and 53 percent longer to transit the Arctic Ocean, depending on exactly which waters Russia denied access to.

To conclude the discussion of this factor, it is important to note that the Arctic is an area of particular uncertainty with respect to law and its interpretation, which are still in their infancy because they were of little interest until fairly recently. Further, the melting of the ice boundary that traditionally protected Russia puts it in a new security configuration that might alter its incentive to abide by the law. Pre-

---

[66] Canada, too, claims that the Northwest Passage is internal waters, allowing it to control transit passage. The United States is of the view that both the Northern Sea Route and the Northwest Passage are international straits, for which UNCLOS prescribes a right to "transit passage" (See UNCLOS, Part III, Section 2: Transit Passage, 1982). Vice Admiral James Houck (Ret.), *The Opportunity Costs of Ignoring the Law of Sea Convention in the Arctic*, Hoover Institution, February 19, 2014.

[67] Zysk and Titley suggest such a use by China of Canada's claim that the Northwest Passage is internal waters. Zysk and Titley, 2015, p. 176.

[68] In fact, a maritime route over the North Pole could be more accessible than the Northwest Passage in some future climatic scenarios because of the persistent ice in parts of Canada and Greenland previously discussed.

[69] This analysis imagines a vessel transiting between Navrik and the Bering Strait, by way of example.

viously, this ice protection and Russia's access capabilities—with five times the number of icebreakers as the next Arctic country—ensured that its influence in the region was largely accepted, no matter what the law technically said. With less of an ice boundary, Russia may need different means to assert its status in the region.

Russia's Perception of a Threat from NATO in the Arctic

Could there be a circumstance in the future under which Russia might react in the Arctic as suddenly and aggressively as it did in Ukraine? The fourth factor we discuss relates to the possibility that Russia may come to see an immediate threat in the Arctic and respond by military means. Although there are ". . . a wide spectrum of potential challenges and threats to the security of Russia . . . being formed in the Arctic," as Russian Vice Minister of Defense Nikolay Pankov put it in 2015,[70] we focus on NATO in particular because of a potential spill-over of current tensions in eastern Europe and around the Baltic, as well as evidence that Russian leadership is fearful of a Northern encirclement of Russia. We first examine why the Kremlin would pursue a potentially costly confrontation in a region that, as we have argued in our discussion of the other factors, presents very limited reasons (if any) to raise geopolitical tensions. We then turn to some scenarios related to NATO that Russia could employ to justify military response to a threat in the region.

Domestic politics are an important factor that may play a role in whether Russia would react to a perceived provocation or potential threat to its territory. Russia's Arctic covers more than half of Russia's territory, and provides 20 percent of the nation's gross domestic product.[71] The Soviet Union invested massively in the region and, after the economic difficulties of the 1990s were partly overcome, Putin

---

[70] "Минобороны: угроза безопасности РФ формируется в Арктике [Ministry of Defense: The Threat to Russian Security Is Being Formed in the Arctic]," *Izvestia*, November 25, 2015.

[71] Wilson Rowe and Blakkisrud, 2014, p. 68.

renewed large-scale investments in the Arctic and its use domestically as a marker of Russian national identity.[72] As a result, the Arctic is viewed in Russia as an important domestic issue before it is viewed through an international or diplomatic lens.[73] Russia's Arctic could therefore be used as a nationalistic stake to shore up domestic support, particularly in times of political and economic difficulties.[74]

Several elements suggest that such a strategy is plausible. First, the literature on the "rally-around-the-flag" effect supports the notion that external conflicts boost domestic support for political leaders, at least in the short term.[75] Putin experienced this effect after the Ukraine crisis, when his popularity reached its highest level ever in June 2015, with an 89-percent approval rate—up almost 30 points from 2013.[76] A second factor is Putin's focus on regime survival.[77] He is particularly wary of Russia being the location of another "color revolution," a concern made more immediate by a series of mass street protests in 2011–2012. This issue figures prominently in Russia's 2015 National Security Strategy and, according to some analysts, was one of the key reasons Russia intervened in Ukraine.[78] In this regard,

---

[72] Laruelle, 2014, pp. 9–12; Koivurova, Käpylä, and Mikola, 2015, pp. 5–6; Wilson Rowe and Blakkisrud, 2014, p. 68.

[73] The above-mentioned analysis of media coverage of the Arctic by *Rossiiskaya gazeta* shows a dominance of domestic themes (in comparison with the following other themes: security; shipping; research; climate; energy; and official statements and documents). Wilson Rowe and Blakkisrud, 2014, p. 75.

[74] Koivurova, Käpylä, and Mikola, 2015, p. 5; Rahbek-Clemmensen, 2015, p. 337.

[75] On the diversionary theory of international conflict, see John Mueller, "Presidential Popularity from Truman to Johnson," *American Political Science Review*, Vol. 64, No. 1, March 1970; Jong R. Lee, "Rallying Around the Flag: Foreign Policy Events and Presidential Popularity," *Presidential Studies Quarterly*, Vol. 7, No. 4, Fall 1977; T. Clifton Morgan and Christopher J. Anderson, "Domestic Support and Diversionary External Conflict in Great Britain, 1950–1992," *Journal of Politics*, Vol. 61, No. 3, August 1999.

[76] Alberto Nardelli, Jennifer Rankin, and George Arnett, "Vladimir Putin's Approval Rating at Record Levels," *The Guardian*, July 23, 2015.

[77] See, for instance, Oliker et al., 2015, p. 22; Fiona Hill, "Rocky Times Ahead for Obama and Putin," Brookings, website, November 13, 2012.

[78] Nicolas Bouchet, "Russia's 'Militarization' of Colour Revolutions," *Policy Perspectives*, Center for Security Studies/ETH Zurich, Vol. 4, No. 2, January 2016, p. 1; Oliker et al.,

the economic recession in Russia, borne out of Western sanctions in addition to low hydrocarbon prices and capital flight, increases pressure on the Kremlin. A third factor is Putin's "social contract" with the Russian population that has been in place since his return to the presidency in 2012, promising a great power status for Russia again in exchange for a limitation of individual freedoms.[79] Achieving that great power status in the eye of public opinion might involve taking a more assertive stance in the Arctic, particularly if it is in response to a perceived foreign provocation from NATO. A fourth factor is the fact that the West—or, as seems increasingly likely, the European Union[80]—lifting sanctions against Moscow actually could increase internal pressure on Putin, who has found it convenient to blame Russia's economic woes on foreign hostility and may have to look for another rallying theme.[81]

The likelihood of Putin using the Arctic in such a way seems low, given there is little evidence that public opinion has shaped his foreign policy so far.[82] However, public opinion has contributed to shaping some elements of domestic politics—and the Arctic,[83] being so closely integrated with Russia's identity, as well as its economic and military survival, blurs the line between foreign and domestic realms.[84] Russia's "defense" of its Arctic region could therefore be used for domestic pur-

---

2015, p. 23.

[79] Alexander Baunov, "Ever So Great: The Dangers of Russia's New Social Contract," Carnegie Endowment for International Peace, website, June 15, 2015; conversation with Arctic experts, Copenhagen, January 2016.

[80] See, for instance, James Kanter, "E.U. to Extend Sanctions Against Russia, but Divisions Show," *New York Times*, December 18, 2015; Robin Emmott and Gabriela Baczynska, "Italy, Hungary Say No Automatic Renewal of Russia Sanctions," Reuters, March 14, 2016.

[81] Michael Birnbaum, "A Year into a Conflict with Russia, Are Sanctions Working?" *Washington Post*, March 27, 2015. On the factors behind Russia's economic recession, see Marek Dabrowski, "The Systemic Roots of Russia's Recession," *Bruegel Policy Contributions*, Vol. 2015, No. 15, Brussels, Belgium, October 2015.

[82] Jeffrey Mankoff, *Russian Foreign Policy: The Return of Great Power Politics*, 2nd ed., Lanham, Md.: Rowman & Littlefield Publishers, 2012, p. 61; Oliker et al., 2015, p. 16.

[83] Oliker et al., 2015, p. 16.

[84] Laruelle, 2014, p. 24.

poses, particularly if internal discontent rose to the point where Putin feared for the survival of his regime. Such justification would likely be used in combination with the perception, by both the Kremlin and domestic audiences, of an external threat, lest Putin risk having the population balk at a costly and senseless confrontation with the West at a time of economic difficulties—potentially resulting in precisely the type of internal contestation that the regime most wants to avoid. NATO presents a plausible threat that, from Russia's perspective, could justify military intervention in the Arctic.

There are a number of ways in which Russia could claim that NATO displays an aggressive intent in the Arctic; for example, increased military presence of NATO members, or greater NATO involvement as an organization in the region. Since the mid-2000s, all Arctic nations have undertaken to modernize or augment their Arctic-relevant military assets, not unlike what Russia was doing at the same time.[85] In 2009, Norway moved its National Joint Headquarters from the south of the country to Bodø in the North, and invested in new frigates and F-35 Joint Strike Fighters. Denmark established a joint Arctic Command in Nuuk (Greenland) and procured two patrol vessels and four frigates that can all operate in first-year ice.[86] Although their efforts do not compare with Russia's in terms of scale, they have also included large-scale exercises—such as Norway's March 2015 exercises in Finnmark, the region bordering Russia, which was also its largest winter exercise in that region since 1967[87]—and a higher degree of military cooperation. Sweden, for instance, promotes the creation of a standby Nordic-Baltic Battle Group within Nordic Defense Cooperation (NORDEFCO).[88] In March 2015, Denmark, Finland,

---

85 Depledge, 2015, p. 59.

86 Le Mière and Mazo, 2013, pp. 88–89.

87 Thomas Nilsen, "Norway Launches High North Military Exercise," *Barents Observer*, March 9, 2015.

88 Gerard O'Dwyer, "Sweden Proposes Aggressive Nordic Defense," *Defense News*, February 10, 2015a. NORDEFCO's 2015 annual report notes that

> We have to take account of the actions taken by Russia and not the Kremlin's rhetoric. Russia is making substantial investments in her Armed forces, with the aim of enhancing

Iceland, Norway, and Sweden signed an agreement to increase their defense cooperation (and to stand in solidarity with the Baltic states) that they openly described as a reaction to Russia's actions.[89] Russia has responded in a tit-for-tat manner, as when it carried out large-scale maneuvers in May 2015 in an apparent response to the biannual Arctic Challenge exercise carried out by Nordic nations.[90] It remains to be seen how it will respond to the United States' intention to refurbish the Keflavik Naval Air Station in Iceland, which had been closed in 2006.[91]

While Russia has long tried to discourage NATO from getting more involved in the Arctic,[92] this issue has also divided those Arctic nations that are NATO members. While Norway favors heavier NATO involvement, at least in terms of preparedness and exercises, Canada and Denmark have been far more reluctant to see non-Arctic nations involved in the region.[93] As a result, NATO's interest in the Arctic has remained limited. The region is not mentioned in the Strategic Concept that NATO adopted in 2010 at the Lisbon Summit, and in 2013 the organization had made clear that it had no intention of greater involve-

---

> its military capabilities, and has demonstrated a will to apply military means to achieve political goals, even when it violates principles of international law. Russia has increased her military exercises and intelligence operations in the Baltic Sea region and the High North.

Swedish Ministry of Defense, *NORDEFCO Annual Report 2015*, January 2016, p. 32.

[89] Balazs Koranyi and Terje Solsvik, "Nordic Nations Agree on Defense Cooperation Against Russia," Reuters, April 9, 2015.

[90] Sergunin and Konyshev, 2015, p. 404; Gerard O'Dwyer, "Tensions High as Russia Responds to Exercise," *Defense News*, May 31, 2015b; "Arctic Challenge Exercise," Norwegian Ministry of Defense website, June 13, 2015.

[91] Kristina Lindborg, "Why Does the Pentagon Want to Refurbish a Base in Iceland?" *Christian Science Monitor*, March 27, 2016; Gregory Winger and Gustav Petursson, "Return to Keflavik Station: Iceland's Cold War Legacy Reappraised," *Foreign Affairs*, February 24, 2016.

[92] Laruelle, 2014, p. 14; Klimenko, 2016, p. 15.

[93] Brooke A. Smith-Windsor, *Putting the 'N' Back into NATO: A High North Policy Framework for the Atlantic Alliance?* NATO Defense College Research Division, No. 94, July 2013, p. 5; Le Mière and Mazo, 2013, pp. 125–126; Helga Haftendorn, "NATO and the Arctic: Is the Atlantic Alliance a Cold War Relic in a Peaceful Region Now Faced With Non-Military Challenges?" *European Security*, Vol. 20, No. 3, September 2011, pp. 341–342.

ment.[94] While the Ukraine crisis has not directly brought new NATO interest to the Arctic, it has increased tension between Russia and its Arctic partners—particularly those NATO members with access to the Baltic Sea. In March 2015, for instance, the Russian ambassador to Denmark stated that Danish frigates could be targeted by Russia if they joined NATO's missile shield.[95]

If Russia were to perceive a substantially greater NATO presence near its Arctic region, it could choose to respond by testing the organization's resolve and unity, possibly with a minor encroachment in Svalbard or even northern Norway. This latter contingency was highlighted in a 2015 expert report commissioned by the Government of Norway, which underlined that in the event of a crisis, Russia might secure the Kola Peninsula—the main location of its nuclear deterrent—by gaining control of adjacent areas, including parts of northern Norway, the Barents Sea, and the Norwegian Sea.[96] While this would certainly be a bolder Russian move against NATO than contesting rights of two of the Alliance's members to the Arctic seabed (as described earlier), it is unclear whether it would trigger a NATO military response or be considered under the threshold for such an action. Here again, the NAC might have a difficult time reaching a consensus. Failure to react, however, could be damaging for NATO—and a success in itself for Russia—because it would demonstrate that Norway does not stand to benefit much from the organization—a contingency for which Norway has been preparing through the concept of "threshold defense," the ability to defend the country in instances where aggression is considered under the threshold for intervention.[97] Not only would this undermine the principle of collective security on which NATO relies, it would also deter potential newcomers, such as Sweden and Finland, from ever joining.

---

[94] Smith-Windsor, 2013, p. 1; NATO, *Active Engagement, Modern Defence: Strategic Concept for the Defence and Security of the Members of the North Atlantic Treaty Organisation*, Lisbon, November 19, 2010.

[95] Julian Isherwood, "Russia Warns Denmark its Warships Could Become Nuclear Targets," *The Telegraph*, March 21, 2015.

[96] Staalesen, 2015.

[97] Depledge, 2015, p. 62 and note 2.

Should Sweden and Finland choose to join NATO, this would likely activate Russia's fear of encirclement and perception of a threat. Russia has already warned that it would react to such a move when Russian Foreign Minister Lavrov declared to a Swedish newspaper that "If military infrastructure draws close to Russian borders, we will naturally take the necessary technical-military measures."[98] Similarly, Russia has made clear to Finland that it would consider a Finnish membership to NATO as an offensive measure, with Putin's personal envoy Sergei Markov declaring in an interview in June 2014 that such a move could start World War III.[99]

Sweden and Finland have come closer to NATO over the years. At the organization's 2014 Wales Summit, both countries integrated the Enhanced Opportunities Partners program, which will increase their level of dialogue and cooperation with NATO.[100] They also signed a host-nation agreement that will facilitate their hosting of NATO forces for training and exercises.[101] They already take part regularly in NATO exercises, such as the annual BALTOPS exercise.[102] Yet the question of membership remains deeply divisive in both countries. In Sweden, left-wing and far-right parties remain opposed to joining NATO,[103] even

---

98 "Finland Risks 'Serious Crisis' With Russia if it Joints NATO, Experts Warn," Agence France-Presse, April 30, 2016.

99 Thomas Nilsen, "Putin Envoy Warns Finland Against Joining NATO," *Barents Observer*, June 9, 2014.

100 Finland and Sweden were two of five countries to integrate this Enhanced Opportunities Partners Program. The others are Australia, Georgia, and Jordan. NATO, "NATO Secretary General Welcomes Deepening Cooperation and Dialogue with Finland and Sweden," web page, December 1, 2015b.

101 Supreme Headquarters Allied Powers Europe Public Affairs Office, "Finland and Sweden Sign Memorandum of Understanding with NATO," NATO, September 5, 2014; Gerard O'Dwyer, "Sweden and Finland Pursue 'Special Relationship' With NATO," *Defense News*, October 10, 2014.

102 NATO, "BALTOPS 16," NATO Naval Striking and Support Forces website, undated; NATO, "NATO Allies Begin Naval Exercise BALTOPS in the Baltic Sea," web page, June 20, 2015a.

103 Gerard O'Dwyer, "New Poll Shows Sharp Shift in NATO Support," *Defense News*, September 17, 2015c.

as public support has been increasing over the years with those in favor of membership outnumbering their opponents for the first time in October 2014, when 37 percent of surveyed individuals were in favor of membership and 36 percent were against.[104] In September 2015, support for membership reached 41 percent.[105] However, a new shift in mid-2016, with only 33 percent expressing support for membership and 49 percent opposing it, suggests that Swedish opinion on the matter is still very much in flux.[106] Finland's public opinion on NATO membership is even more ambivalent, with only 27 percent of respondents in a February 2015 survey expressing support.[107] In April 2016, an expert report commissioned by the Finnish government to analyze potential implications of NATO membership for the country came to the conclusion that "Membership would probably also lead to a serious crisis with Russia, for an undefined period of time."[108]

The two countries' prospects for NATO membership are closely tied. The Finnish expert report highlighted that country's increased vulnerability if Sweden were to join NATO alone, leaving Finland as the only non-NATO country in northern Europe.[109] Overall, given the mixed support domestically and both countries' reluctance to break with their history of neutrality, any decisive move toward a NATO membership appears unlikely in the near future, even though it will remain high on their respective political agendas.

---

[104] Johan Ahlander, "Poll Shows More Swedes in Favor of NATO for First Time," Reuters, October 29, 2014.

[105] O'Dwyer, 2015c.

[106] Gabriela Baczynska, "Wary of Russia, Sweden and Finland Sit at NATO Top Table," Reuters, July 8, 2016, citing figures from a SvD/SIFO opinion poll.

[107] Juha-Pekka Raeste, "HS-gallup: Enemmistö suomalaisista vastustaa yhä Nato-jäsenyyttä [The Majority of Finns Still Oppose NATO Membership]," *Helsingin Sanomat*, March 5, 2015.

[108] "Finland Risks 'Serious Crisis' With Russia if it Joints NATO, Experts Warn," 2016.

[109] "Finland Risks 'Serious Crisis' With Russia if it Joints NATO, Experts Warn," 2016.

CHAPTER FOUR

Conclusion and Policy Implications

This chapter outlines our main research findings and the policy implications for the United States. We further examine these policy implications in light of the United States' current Arctic Strategy to identify whether a full implementation of the strategy—which was not achieved as of late 2016—might address some of these concerns.

Findings

Our first two research questions focused on the factors that have maintained the Arctic as an area of cooperation, and their ability to sustain such cooperation in the face of dramatic changes that will likely take place in the Arctic. They produced five key findings:

1. **Russia's current militarization of its Arctic region does not, in itself, suggest increased potential for conflict, with the exception of accidental escalation.** Russia is still a long way from reestablishing Cold War levels of military presence in the Arctic, and is unlikely to use Arctic-based assets effectively in other, more likely, contingencies—for instance, in the Baltics. Yet increased military presence—not just from Russia but also other Arctic countries—increases risks of collisions and accidental escalation.
2. **Russia's cooperative stance in the Arctic cannot be taken for granted.** Anticipating Russia's behavior in the Arctic raises the question of whether its intentions can be confidently inferred

from past behavior. Russia's intervention in Ukraine has been widely described as a surprise—even seasoned observers did not expect the Maidan protests in Kyiv to trigger such an aggressive move from Russia. In the Arctic, the number of mechanisms (e.g., agreements, diplomatic organizations) through which Russia cooperates on Arctic affairs could make it difficult to abandon this stance in rapid fashion. Yet, Russia has shown some unpredictable behavior in that region as well. For instance, it reacted very mildly to Greenpeace activists boarding the Prirazlomnaya oil rig in August 2012, but arrested the crew and impounded their ship in a similar incident the following year.[1] This, in addition to its mix of cooperative and assertive rhetoric on the Arctic, makes Russia's intentions particularly difficult to read.

Furthermore, there are no factors that make it inherently beneficial for Russia to cooperate on Arctic issues. While destabilizing the region would limit Russia's potential to benefit from its Arctic resources, which its national priorities clearly indicate it wishes to do, our analysis of the second factor (increased interest in Arctic resources) suggests that even economic factors will not necessarily steer Russia toward cooperation in the future. If economic ambitions grow increasingly out of reach—for instance, because of low hydrocarbon prices, capital flight, and/or the loss of foreign investment and expertise—Russia could have less of an incentive to cooperate and might engage instead in inflammatory actions and rhetoric.

3. **Sea ice decline projections suggest that while Russia's northern shore will be increasingly exposed, increased maritime access overall (including through a trans-Arctic route in the long term) will reduce Russia's ability to control Arctic shipping lanes or block them in the event of a conflict.** From Russia's viewpoint, this both increases its perceived vulnerabil-

[1] Dmitry Gorenburg, "Russian Interests and Policies in the Arctic," *War on the Rocks*, blog, August 7, 2014b; Trude Pettersen, "Greenpeace Occupying Prirazlomnaya Platform," *Barents Observer*, August 24, 2012a; Gorenburg, 2014a.

ity to potential attacks and removes what might have been a powerful tool of coercion against other countries in the event of a conflict. As a result, Russia will likely continue to militarize the Arctic in the medium to long term, if only to protect its strategic assets and infrastructure in the region.

4. **While Russia has mostly benefited from UNCLOS decisions in the past, there would be nothing to stop it from ignoring or distorting UNCLOS recommendations if it judged such recommendations contrary to its interests.** It is worth noting that the UNCLOS decision itself bears little risk of conflict, at least in the short term. The rights it would recognize would not lead to actual resource exploitation for years, possibly decades. In the case of a negative outcome for its claim, Russia could also resubmit and simply prolong the process.
5. **Russia would likely feel threatened by an expansion of NATO's role in the Arctic.** The Kremlin has shown consistent hostility to increased support for NATO in Sweden and Finland, and to a larger NATO influence in the region, suggesting that keeping NATO at bay is a solid, and permanent, tenet of its Arctic policy. Whether this would lead to conflict—for instance, with Russia testing the alliance's commitment through minor encroachments on Arctic members' territory—is doubtful, unless such a war also served domestic purposes for Putin and supported his hold on power.

Policy Implications

Our third research question focused on options for U.S. policy to help mitigate the effects of the previously outlined factors and to contain tensions. The fact that Russia's behavior in the Arctic could change from cooperative to conflictual and is difficult to foresee (Finding 2) warrants close U.S. attention to the region and careful observation of developments in the Arctic—not just the types and numbers of military assets positioned by Russia, but also how the Arctic is portrayed in strategy and policy, when the Kremlin is choosing to cooperate on

Arctic affairs, and what investments are being planned or made in energy and other civilian or multipurpose infrastructure. Monitoring of the region may require encouraging improvements in Arctic region domain awareness and access, through continuing and expanding, as necessary, funding for

- mapping (including of underwater topography)
- vessels and aircraft that can operate in Arctic conditions[2]
- maintaining existing infrastructure and assets
- development of multipurpose ports and airstrips that can facilitate access[3]
- enhancing communications systems to promote a safe operating environment and help avoid unintended conflict
- further allocating intelligence, surveillance, and reconnaissance assets that can help increase the transparency of foreign Arctic activities to help prevent misunderstandings that can lead to conflict.

The U.S. Arctic Strategy issued in 2013 includes "Enhance Arctic Domain Awareness" as an element of its first line of effort entitled "Advance United States Security Interests." This focus area is further developed in the 2014 Implementation Plan for the National Strategy for the Arctic Region.[4] Due to the high cost of many of the assets we have discussed, the decision to implement this part of the strategy will largely depend on Congress.

Unpredictability also suggests that special care should be taken to avoid accidental escalation of small-scale incidents (Finding 1). This can be done through supporting activities that bring the United States and Russia together on Arctic issues—for instance, through institutions such as the Arctic Council, the Arctic Coast Guard Forum, and the International Maritime Organization; joint activities, such as

---

[2] Requirements will have to be developed to express specific needs.

[3] Specific needs will have to be identified first.

[4] White House, 2013, p. 2; White House, *Implementation Plan for the National Strategy for the Arctic Region*, Washington, D.C., January 2014, pp. 7–8.

safety and environmental exercises, collaborative scientific research; and information-sharing, such as data related to commercial shipping traffic. It could also be done by reducing Department of Defense barriers to participating in international Arctic activities that involve Russia when the focus is military support to civil authorities (such as SAR exercises). Familiarity with the Russian officials and organizations that are concerned with Arctic matters help ensure safe and transparent operations in the region, and may prevent small incidents from turning into larger ones.

While "Strengthen International Cooperation" represents an entire line of effort of the U.S. Arctic Strategy, it does not mention creating a forum dedicated to security issues—an initiative that could promote coordination, facilitate information-sharing, reduce uncertainty, and possibly help limit the potential for unintended escalation of tensions. Given present restrictions and tensions, a good way to include Russia in Arctic security conversations could be to leverage existing international conferences, such as that run by the Arctic Circle Assembly, where participation can be ad hoc and security is already an item of discussion.

Russia's increased vulnerability on its northern shore (Finding 3) and sensitivity to an increased NATO presence (Finding 5) in the Arctic region writ large also suggest that even limited NATO incursions for routine activities (such as exercises) have the potential to fuel tensions when seen against the background of stronger support for NATO on the part of Sweden and Finland. While this does not mean that NATO should halt its activities in the region, it suggests the necessity to strike a balance between ensuring that NATO has some capability and experience to support Arctic operations without establishing a presence in the region that would create tensions between Arctic nations, and particularly with Russia. This includes supporting measures designed to strengthen NATO's ability to conduct operations in cold-weather conditions through training and exercises (such as Cold Response), and pursue efforts started at the 2014 Wales Summit and confirmed at the 2016 Warsaw Summit to adapt to the new threat environment, such as improving and speeding up decisionmaking processes within the NAC and improving sharing processes for intelligence assessments. Lessons learned from the employment of NATO's new capabilities in the Baltics could also inform planning for potential Arctic contingencies. A new Arctic security forum as described above would also be instrumental in reducing tensions by ensuring that NATO does not become the only forum of discussion of Arctic issues—a forum from which Russia is excluded.

Finally, the United States would be in a better position to pressure Russia to abide by its commitment to UNCLOS—whether Russia denies the CLCS's eventual decision or distorts it through its own interpretation of the law (Finding 4)—if it were an UNCLOS signatory itself. Ratifying UNCLOS would ensure that the United States has the appropriate foundation—like the other Arctic countries and many others around the world—to help resolve relevant maritime disputes according to a cooperative process. This step is mentioned in the U.S. Arctic Strategy as an element of the third line of U.S. effort in the Arctic ("Strengthen International Cooperation"), noting that "Accession to the Convention would protect U.S. rights, freedoms, and uses of the sea and airspace throughout the Arctic region, and strengthen our arguments for freedom of navigation and overflight through the Northwest Passage and the Northern Sea Route."[5]

Although we recognize that there are substantial barriers to fully addressing these policy implications because of political, budgetary, and/or other challenges, it is nonetheless important to highlight them because they are fundamental to U.S. Arctic strategy and continued cooperation with Russia in the region. Failing to prepare for these transformations may have serious implications for some key priorities of the United States, such as promoting freedom of navigation; ensuring the safety and environmental security of U.S. citizens living in the Arctic; and maintaining domain awareness in a region that may become both increasingly militarized and economically significant.

[5] White House, 2013, p. 9.

Acknowledgments

The authors are grateful for the support and help of many individuals. In particular, we would like to thank the government officials and Arctic researchers—mainly in Washington, D.C.; Oslo, Norway; Tromsø, Norway; and Copenhagen, Denmark—who took the time to share their expertise with us. We also thank U.S. Northern Command for extending an invitation to us to attend the 2016 Arctic Collaborative Workshop in Fairbanks, Alaska, and we are indebted to the participants of the May 2016 RAND Arctic Roundtable, who provided us with valuable insights on Arctic security and prospects.

We greatly appreciate the comments of David P. Auerswald of the National War College, Anika Binnendijk of RAND, and Lawson Brigham of the University of Alaska Fairbanks, who reviewed the initial draft of this report, significantly improving its quality and accuracy.

We are further indebted to our RAND colleagues who provided research and editing support, as well as valuable advice, over the course of this project, particularly Katherine Anania, Christina Bartol Burnett, Scott Boston, Christopher S. Chivvis, Dara Massicot, Michael J. McNerney, Katya Migacheva (whom we also thank for her translations of Russian sources), David Ochmanek, Clinton Reach, Constantine Samaras, Scott Savitz, and Timothy Smith. Howard J. Shatz and Susan L. Marquis provided invaluable support throughout the research and publication process.

Responsibility for the content of this report lies solely with the authors.

Abbreviations

ATAM	Arctic Transit Accessibility Model
BALTOPS	Baltic operations
CLCS	United Nations Commission on the Limits of the Continental Shelf
EEZ	exclusive economic zone
NAC	North Atlantic Council
NATO	North Atlantic Treaty Organization
nm	nautical mile
SAR	search and rescue
UN	United Nations
UNCLOS	United Nations Convention on the Law of the Sea
USGS	U.S. Geological Survey

References

Ahlander, Johan, "Poll Shows More Swedes in Favor of NATO for First Time," Reuters, October 29, 2014. As of June 7, 2016:
http://www.reuters.com/article/2014/10/29/us-sweden-nato-idUSKBN0II1XN20141029

Ahmari, Sohrab, "The New Cold War's Arctic Front," *Wall Street Journal*, June 9, 2015. As of June 3, 2016:
http://www.wsj.com/articles/the-new-cold-wars-arctic-front-1433872323

Aksenov, Yevgeny, Ekaterina E. Popova, Andrew Yool, A. J. George Nurser, Timothy D. Williams, Laurent Bertino, and Jon Bergh, "On the Future Navigability of Arctic Sea Routes: High-Resolution Projections of the Arctic Ocean and Sea Ice," *Marine Policy*, Vol. 75, January 2017, pp. 300–317. As of January 18, 2017:
http://www.sciencedirect.com/science/article/pii/S0308597X16000038

"Arctic Challenge Exercise," Norwegian Ministry of Defense website, June 13, 2015. As of June 7, 2016:
https://forsvaret.no/en/exercise-and-operations/exercises/ace

Arctic Council, "Arctic Economic Council," web page, September 1, 2015. As of June 6, 2016:
http://www.arctic-council.org/index.php/en/our-work2/8-news-and-events/195-aec-2

Atland, Kristian, "Russia's Armed Forces and the Arctic: All Quiet on the Northern Front?" *Contemporary Security Policy*, Vol. 32, No. 2, August 2011, pp. 267–285.

Baczynska, Gabriela, "Wary of Russia, Sweden and Finland Sit at NATO Top Table," Reuters, July 8, 2016. As of January 24, 2017:
http://www.reuters.com/article/us-nato-summit-nordics-idUSKCN0ZO1EO

Baev, Pavel K., "Russia's Arctic Ambitions and Anxieties," *Current History*, October 2013, pp. 265–270.

Baunov, Alexander, "Ever So Great: The Dangers of Russia's New Social Contract," Carnegie Endowment for International Peace, website, June 15, 2015. As of June 7, 2016:
http://carnegieendowment.org/2015/06/15/ever-so-great-dangers-of-russia-s-new-social-contract/ialt

Birnbaum, Michael, "A Year into a Conflict with Russia, Are Sanctions Working?" *Washington Post*, March 27, 2015. As of June 7, 2016:
https://www.washingtonpost.com/world/europe/a-year-into-a-conflict-with-russia-are-sanctions-working/2015/03/26/45ec04b2-c73c-11e4-bea5-b893e7ac3fb3_story.html

Bodner, Matthew, "Russia's Polar Pivot: Moscow Revamps, Re-Opens Former Soviet Bases to Claim Territories," *Defense News*, March 11, 2015a.

———, "New Russian Naval Doctrine Enshrines Confrontation with NATO," *Moscow Times*, July 27, 2015b. As of June 3, 2016:
http://www.themoscowtimes.com/business/article/new-russian-naval-doctrine-enshrines-confrontation-with-nato/526277.html

Boersma, Tim, and Kevin Foley, *The Greenland Gold Rush: Promises and Pitfalls of Greenland's Energy and Mineral Resources*, Washington, D.C.: Brookings Institution, September 2014.

Borgerson, Scott G., "Arctic Meltdown: The Economic and Security Implications of Global Warming," *Foreign Affairs*, Vol. 87, No. 2, 2008, pp. 63–77.

Bouchet, Nicolas, "Russia's 'Militarization' of Colour Revolutions," *Policy Perspectives*, Center for Security Studies/ETH Zurich, Vol. 4, No. 2, January 2016. As of June 7, 2016:
https://www.ethz.ch/content/dam/ethz/special-interest/gess/cis/center-for-securities-studies/pdfs/PP4-2.pdf

Byers, Michael, *Who Owns the Arctic? Understanding Sovereignty Disputes in the North*, Madeira Park (BC), Canada: Douglas and McIntyre Ltd., 2009.

Carlsson, Märta, and Niklas Granholm, *Russia and the Arctic: Analysis and Discussion of Russian Strategies*, Stockholm, Sweden: Swedish Defence Research Agency, March 2013.

Central Intelligence Agency, "Political Arctic Region," *World Factbook*, undated. As of January 20, 2017:
https://www.cia.gov/library/publications/the-world-factbook/docs/refmaps.html

Chivers, C. J., "Russians Plant Flag on the Arctic Seabed," *New York Times*, August 3, 2007. As of June 3, 2016:
http://www.nytimes.com/2007/08/03/world/europe/03arctic.html?_r=0

Christiansson, Magnus, *Strategic Surprise in the Ukraine Crisis: Agendas, Expectations, and Organizational Dynamics in the EU Eastern Partnership Until the Annexation of Crimea 2014*, thesis, Stockholm, Sweden: Swedish National Defence College, August 2014.

Conley, Heather A., "Russia's Influence on Europe," in Craig C. Cohen and Josiane Gabel, eds., *2015 Global Forecast: Crisis and Opportunity*, Washington, D.C.: Center for Strategic and International Studies, 2014.

Conley, Heather A., David Pumphrey, Terrance M. Toland, and Mihaela David, *Arctic Economics in the 21st Century: The Benefits and Costs of Cold*, Washington, D.C.: Center for Strategic and International Studies, July 2013.

Conley Heather A., and Caroline Rohloff, *The New Ice Curtain: Russia's Strategic Reach to the Arctic*, Washington, D.C.: Center for Strategic and International Studies, August 2015.

Dabrowski, Marek, "The Systemic Roots of Russia's Recession," *Bruegel Policy Contributions*, Vol. 2015, No. 15, Brussels, Belgium, October 2015. As of June 7, 2016:
http://bruegel.org/wp-content/uploads/2015/10/pc_2015_15.pdf

"Denmark Challenges Russia and Canada Over North Pole," BBC, December 15, 2014. As of June 3, 2016:
http://www.bbc.com/news/world-europe-30481309

Depledge, Duncan, "Hard Security Developments," in Juha Jokela, ed., *Arctic Security Matters*, Paris, France: European Union Institute for Security Studies, Report No. 24, June 2015, pp. 59–67.

Dimitrakopoulou, Sophia, and Andrew Liaropoulos, "Russia's National Security Strategy to 2020: A Great Power in the Making?" *Caucasian Review of International Affairs*, Vol. 4, No. 1, Winter 2010.

Dodds, Klaus, "The Arctic: From Frozen Desert to Open Polar Sea?" in Daniel Moran and James A. Russell, eds., *Maritime Strategy and Global Order: Markets, Resources, Security*, Washington, D.C.: Georgetown University Press, 2016, pp. 149–180.

Ebinger, Charles K., and Evie Zambetakis, "The Geopolitics of Arctic Melt," *International Affairs*, Vol. 85, No. 6, November 2009, pp. 1215–1232. As of January 18, 2017:
http://onlinelibrary.wiley.com/doi/10.1111/j.1468-2346.2009.00858.x/abstract?userIsAuthenticated=false&deniedAccessCustomisedMessage=

Eckel, Mike, "Russia Defends North Pole Flag-Planting," Associated Press, August 8, 2007. As of June 7, 2016:
http://www.washingtonpost.com/wp-dyn/content/article/2007/08/07/AR2007080701554_pf.html

Emmerson, Charles, *The Future History of the Arctic*, New York: Public Affairs, 2010.

Emmott, Robin, and Gabriela Baczynska, "Italy, Hungary Say No Automatic Renewal of Russia Sanctions," Reuters, March 14, 2016.

"Finland Risks 'Serious Crisis' With Russia if it Joints NATO, Experts Warn," Agence France-Presse, April 30, 2016.

"Основы государственной политики Российской Федерации в Арктике на период до 2020 года и дальнейшую перспективу [The Fundamentals of the Russian Federation State Policy in the Arctic for the Period Until 2020 and Beyond]," Moscow, Russia: Government of the Russian Federation, September 18, 2008. As of June 7, 2016:
http://government.ru/media/files/A4qP6brLNJ175I40U0K46x4SsKRHGfUO.pdf

Gautier, Donald L., Kenneth J. Bird, Ronald R. Charpentier, Arthur Grantz, David W. Houseknecht, Timothy R. Klett, Thomas E. Moore, Janet K. Pitman, Christopher J. Shenck, John H. Schuenemyer, Kai Sorensen, Marilyn E. Tennyson, Zenon C. Valin, and Craig J. Wandrey, "Assessment of Undiscovered Oil and Gas in the Arctic," *Science*, Vol. 324, No. 5931, 2009, pp. 1175–1179.

Giles, Keir, and Andrew Monaghan, *Russian Military Transformation—Goal in Sight?* Carlisle, Pa.: Strategic Studies Institute, May 2014.

Gorenburg, Dmitry, "How to Understand Russia's Arctic Strategy," *Washington Post*, February 12, 2014a.

———, "Russian Interests and Policies in the Arctic," *War on the Rocks*, blog, August 7, 2014b. As of June 7, 2016:
http://warontherocks.com/2014/08/russian-interests-and-policies-in-the-arctic/

Gorodiloff, Michael, "Will Russia and China Set up a Rare Earth Metals Cartel?" *Russia Direct*, February 10, 2016. As of May 2, 2016:
http://www.russia-direct.org/opinion/
will-russia-and-china-set-rare-earth-metals-cartel

Gray, Bryce, "As Greenland Ramps Up Mining, Who Will Benefit?" *Arctic Deeply*, March 17, 2016. As of May 2, 2016:
https://www.newsdeeply.com/arctic/articles/2016/03/17/
as-greenland-ramps-up-mining-who-will-benefit

Gressel, Gustav, *Russia's Quiet Military Revolution, and What it Means for Europe*, London: European Council on Foreign Relations, policy brief (ECFR/143), October 12, 2015.

Haftendorn, Helga, "NATO and the Arctic: Is the Atlantic Alliance a Cold War Relic in a Peaceful Region Now Faced With Non-Military Challenges?" *European Security*, Vol. 20, No. 3, September 2011, pp. 337–361.

Herrmann, Victoria, "U.S.-Russian Cooperation in the Arctic," *New York Times*, letter to the editor, May 9, 2016. As of June 7, 2016:
http://www.nytimes.com/2016/05/09/opinion/us-russian-cooperation-in-the-arctic.html?_r=0

Hill, Fiona, "Rocky Times Ahead for Obama and Putin," Brookings, website, November 13, 2012. As of June 7, 2016:
http://www.brookings.edu/research/opinions/2012/11/13-obama-putin-hill

Hoag, Hannah, "Nations Negotiate Fishing in Arctic High Seas," *Arctic Deeply*, April 28, 2016. As of January 19, 2017:
https://www.newsdeeply.com/arctic/articles/2016/04/28/nations-negotiate-fishing-in-arctic-high-seas

Houck, James, Vice Admiral (Ret.), *The Opportunity Costs of Ignoring the Law of Sea Convention in the Arctic*, Hoover Institution, February 19, 2014. As of January 19, 2017:
http://www.hoover.org/research/opportunity-costs-ignoring-law-sea-convention-arctic

Ilulissat Declaration, *Arctic Ocean Conference*, Ilulissat, Greenland, May 27–29, 2008. As of June 3, 2016:
http://www.oceanlaw.org/downloads/arctic/Ilulissat_Declaration.pdf

International Association of Classification Societies, *Requirements Concerning Polar Class*, London, undated. As of January 18, 2017:
http://www.iacs.org.uk/document/public/Publications/Unified_requirements/PDF/UR_I_pdf410.pdf

Isherwood, Julian, "Russia Warns Denmark Its Warships Could Become Nuclear Targets," *The Telegraph*, March 21, 2015. As of June 7, 2016:
http://www.telegraph.co.uk/news/worldnews/europe/denmark/11487509/Russia-warns-Denmark-its-warships-could-become-nuclear-targets.html

Juhasz, Antonia, "Shell is Reeling After Pulling Out of the Arctic," *Newsweek*, October 13, 2015. As of September 22, 2016:
http://www.newsweek.com/2015/10/23/shell-reeling-after-pulling-out-arctic-382551.html

Kanter, James, "E.U. to Extend Sanctions Against Russia, but Divisions Show," *New York Times*, December 18, 2015.

Klimenko, Ekaterina, *Russia's Arctic Security Policy: Still Quiet in the High North?* Solna, Sweden: Stockholm International Peace Research Institute, Policy Paper 45, February 2016.

Knapp, Gunnar, "Arctic Fisheries: Opportunities and Policy Issues," presentation at the UAA Institute of Social and Economic Research, University of Alaska, Anchorage, Alaska, February 27, 2014.

Koivurova, Timo, Juha Käpylä, and Harri Mikola, *Continental Shelf Claims in the Arctic: Will Legal Procedure Survive the Growing Uncertainty?* Helsinki, Finland: Finnish Institute of International Affairs, Briefing Paper 178, August 2015. As of June 3, 2016:
http://www.fiia.fi/en/publication/516/continental_shelf_claims_in_the_arctic/

Koranyi, Balazs, and Terje Solsvik, "Nordic Nations Agree on Defense Cooperation Against Russia," Reuters, April 9, 2015. As of June 7, 2016:
http://www.reuters.com/article/us-nordics-russia-defence-idUSKBN0N02E820150409

Kyle, Kate, "N.W.T.'s Gahcho Kué Diamond Mine Marks Grand Opening Today," CBC News, September 20, 2016. As of January 16, 2017:
http://www.cbc.ca/news/canada/north/gahcho-kue-diamond-mine-official-opening-1.3769779

Laruelle, Marlène, *Russia's Arctic Strategies and the Future of the Far North*, New York: M.E. Sharpe, Inc., 2014.

Le Mière, Christian, and Jeffrey Mazo, *Arctic Opening: Insecurity and Opportunity*, London: International Institute for Strategic Studies, 2013.

Lee, Jong R., "Rallying Around the Flag: Foreign Policy Events and Presidential Popularity," *Presidential Studies Quarterly*, Vol. 7, No. 4, Fall 1977, pp. 252–256.

Lindborg, Kristina, "Why Does the Pentagon Want to Refurbish a Base in Iceland?" *Christian Science Monitor*, March 27, 2016.

Lucas, Edward, *The Coming Storm: Baltic Sea Security Report*, Washington, D.C.: Center for European Policy Analysis, June 2015.

Macalister, Terry, "Climate Change Could Lead to Arctic Conflict, Warns Senior NATO Commander," *The Guardian*, October 11, 2010. As of June 3, 2016:
http://www.theguardian.com/environment/2010/oct/11/nato-conflict-arctic-resources

Mankoff, Jeffrey, *Russian Foreign Policy: The Return of Great Power Politics*, 2nd ed., Lanham, Md.: Rowman & Littlefield Publishers, 2012.

"Морская доктрина Российской Федерации: Владимир Путин провёл совещание, на котором обсуждалась новая редакция Морской доктрины Российской Федерации [Maritime Doctrine of the Russian Federation: Vladimir Putin Held a Conference to Discuss the New Edition of the Maritime Doctrine of the Russian Federation]," transcript, Moscow, Russia: Kremlin.ru, July 26, 2015. As of June 7, 2016:
http://kremlin.ru/events/president/news/50060

Martina, Michael, "China Condemns Russia for Detaining Fishermen," Reuters, July 19, 2012. As of June 7, 2016:
http://www.reuters.com/article/us-china-russia-fishermen-idUSBRE86I0PM20120719

Minister of Indian Affairs and Northern Development and Federal Interlocutor for Métis and Non-Status Indians, *Canada's Northern Strategy: Our North, Our Heritage, Our Future*, Ottowa, 2009.

"Минобороны: угроза безопасности РФ формируется в Арктике [Ministry of Defense: The Threat to Russian Security Is Being Formed in the Arctic]," *Izvestia*, November 25, 2015. As of June 7, 2016:
http://izvestia.ru/news/597206#ixzz43xbZmMhE

Morgan, T. Clifton, and Christopher J. Anderson, "Domestic Support and Diversionary External Conflict in Great Britain, 1950–1992," *Journal of Politics*, Vol. 61, No. 3, August 1999, pp. 799–814.

Mueller, John, "Presidential Popularity from Truman to Johnson," *American Political Science Review*, Vol. 64, No. 1, March 1970, pp. 18–34.

Myers, Steven Lee, "Sea Warming Leads to Ban on Fishing in the Arctic," *New York Times*, July 16, 2015. As of January 19, 2017:
http://www.nytimes.com/2015/07/17/world/europe/sea-warming-leads-to-ban-on-fishing-in-the-arctic.html?_r=0

Nardelli, Alberto, Jennifer Rankin, and George Arnett, "Vladimir Putin's Approval Rating at Record Levels," *The Guardian*, July 23, 2015. As of June 7, 2016:
http://www.theguardian.com/world/datablog/2015/jul/23/vladimir-putins-approval-rating-at-record-levels

NATO—*See* North Atlantic Treaty Organization.

"New Cold War for Resources Looms in Arctic," *Moscow Times*, April 16, 2012. As of June 3, 2016:
http://www.themoscowtimes.com/business/article/new-cold-war-for-resources-looms-in-arctic/456810.html

Nichol, Jim, *Russian Military Reform and Defense Policy*, Washington, D.C.: Congressional Research Service, R42006, August 24, 2011.

Nilsen, Thomas, "Putin Envoy Warns Finland Against Joining NATO," *Barents Observer*, June 9, 2014. As of June 7, 2016:
http://barentsobserver.com/en/security/2014/06/putin-envoy-warns-finland-against-joining-nato-09-06

———, "Norway Launches High North Military Exercise," *Barents Observer*, March 9, 2015. As of June 7, 2016:
http://barentsobserver.com/en/security/2015/03/norway-launches-high-north-military-exercise-09-03

North Atlantic Treaty Organization, "BALTOPS 16," NATO Naval Striking and Support Forces website, undated. As of June 7, 2016:
http://www.sfn.nato.int/activities/current-and-future/exercises/baltops-16.aspx

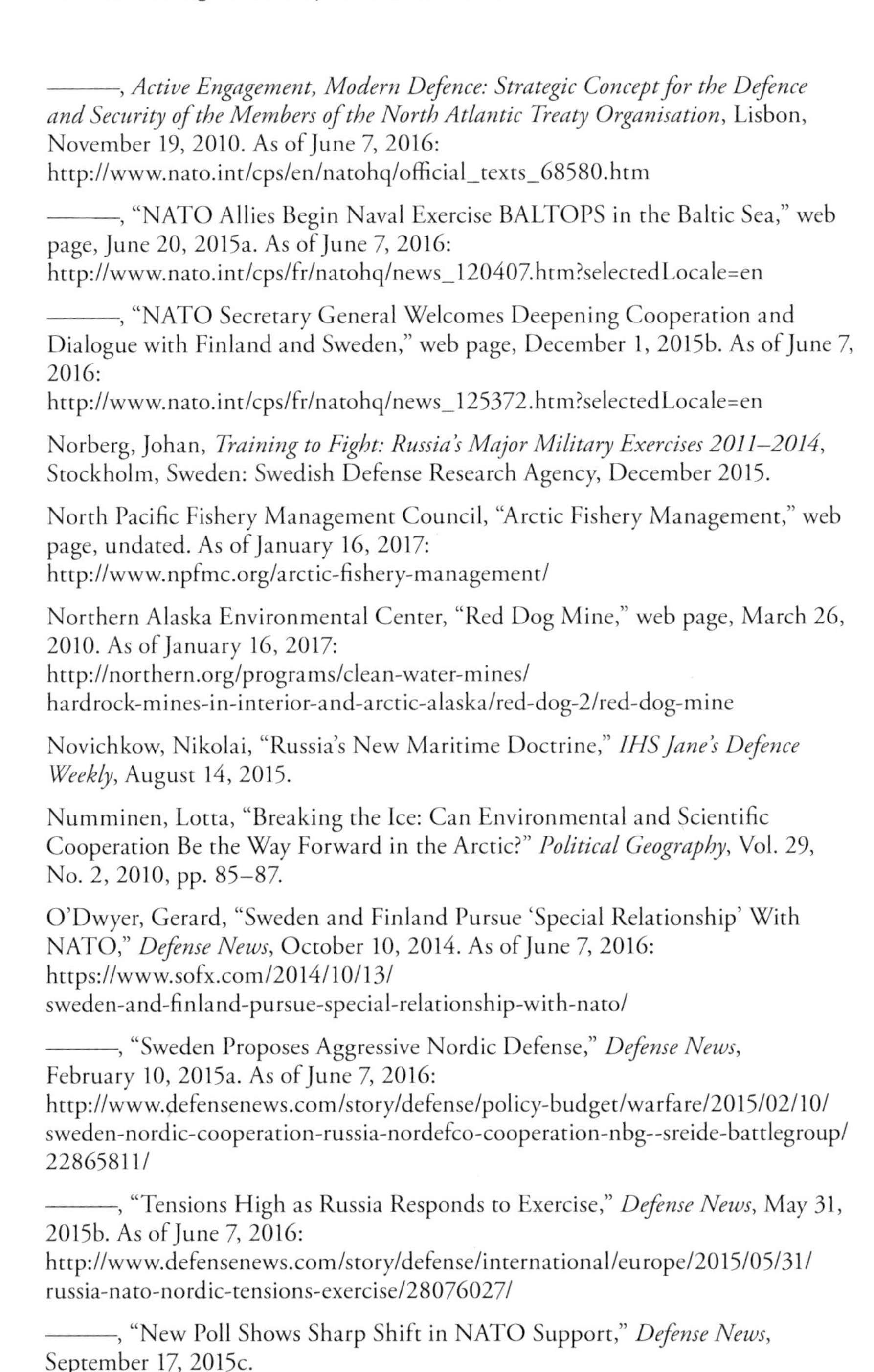

———, *Active Engagement, Modern Defence: Strategic Concept for the Defence and Security of the Members of the North Atlantic Treaty Organisation*, Lisbon, November 19, 2010. As of June 7, 2016:
http://www.nato.int/cps/en/natohq/official_texts_68580.htm

———, "NATO Allies Begin Naval Exercise BALTOPS in the Baltic Sea," web page, June 20, 2015a. As of June 7, 2016:
http://www.nato.int/cps/fr/natohq/news_120407.htm?selectedLocale=en

———, "NATO Secretary General Welcomes Deepening Cooperation and Dialogue with Finland and Sweden," web page, December 1, 2015b. As of June 7, 2016:
http://www.nato.int/cps/fr/natohq/news_125372.htm?selectedLocale=en

Norberg, Johan, *Training to Fight: Russia's Major Military Exercises 2011–2014*, Stockholm, Sweden: Swedish Defense Research Agency, December 2015.

North Pacific Fishery Management Council, "Arctic Fishery Management," web page, undated. As of January 16, 2017:
http://www.npfmc.org/arctic-fishery-management/

Northern Alaska Environmental Center, "Red Dog Mine," web page, March 26, 2010. As of January 16, 2017:
http://northern.org/programs/clean-water-mines/
hardrock-mines-in-interior-and-arctic-alaska/red-dog-2/red-dog-mine

Novichkow, Nikolai, "Russia's New Maritime Doctrine," *IHS Jane's Defence Weekly*, August 14, 2015.

Numminen, Lotta, "Breaking the Ice: Can Environmental and Scientific Cooperation Be the Way Forward in the Arctic?" *Political Geography*, Vol. 29, No. 2, 2010, pp. 85–87.

O'Dwyer, Gerard, "Sweden and Finland Pursue 'Special Relationship' With NATO," *Defense News*, October 10, 2014. As of June 7, 2016:
https://www.sofx.com/2014/10/13/
sweden-and-finland-pursue-special-relationship-with-nato/

———, "Sweden Proposes Aggressive Nordic Defense," *Defense News*, February 10, 2015a. As of June 7, 2016:
http://www.defensenews.com/story/defense/policy-budget/warfare/2015/02/10/
sweden-nordic-cooperation-russia-nordefco-cooperation-nbg--sreide-battlegroup/
22865811/

———, "Tensions High as Russia Responds to Exercise," *Defense News*, May 31, 2015b. As of June 7, 2016:
http://www.defensenews.com/story/defense/international/europe/2015/05/31/
russia-nato-nordic-tensions-exercise/28076027/

———, "New Poll Shows Sharp Shift in NATO Support," *Defense News*, September 17, 2015c.

Oliker, Olga, *Russia's Nuclear Doctrine: What We Know, What We Don't, and What That Means*, Washington, D.C.: Center for Strategic and International Studies, May 2016.

Oliker, Olga, Christopher S. Chivvis, Keith Crane, Olesya Tkacheva, and Scott Boston, *Russian Foreign Policy in Historical and Current Context: A Reassessment*, Santa Monica, Calif.: RAND Corporation, PE-144-A, 2015. As of January 17, 2017: http://www.rand.org/pubs/perspectives/PE144.html

Padrtová, Barbora, "Russian Military Build-Up in the Arctic: Strategic Shift in the Balance of Power or Bellicose Rhetoric Only?" *Arctic Yearbook 2014*, Northern Research Forum and the University of the Arctic Thematic Network on Geopolitics and Security, 2014, pp. 1–19.

Pan, Min, and Henry P. Huntington, "A Precautionary Approach to Fisheries in the Central Arctic Ocean: Policy, Science, and China," *Marine Policy*, Vol. 63, January 2016, pp. 153–157. As of January 19, 2017: http://www.sciencedirect.com/science/article/pii/S0308597X15002997

Panichkin, Ivan, "To Explore and Develop," *Russian International Affairs Council*, November 24, 2015. As of May 17, 2016: http://russiancouncil.ru/en/inner/?id_4=6871#top-content

Parfitt, Tom, "Russia Sends Troops and Missiles to Arctic Bases," *The Times* (UK), December 26, 2015.

Pelyasov, Alexander, "Russian Strategy of the Development of the Arctic Zone and the Provision of National Security Until 2020 (Adopted by the President of the Russian Federation on February 8, 2013, No. Pr-232)," *2013 Arctic Yearbook*, Northern Research Forum and the University of the Arctic Thematic Network on Geopolitics and Security, 2013. As of June 7, 2016: http://www.arcticyearbook.com/commentaries-2013#a10

Pettersen, Trude, "Greenpeace Occupying Prirazlomnaya Platform," *Barents Observer*, August 24, 2012a. As of June 7, 2016: http://barentsobserver.com/en/arctic/greenpeace-occupying-prirazlomnaya-platform-24-08

———, "Russia Sends Mig-31 Interceptors to the Arctic," *Barents Observer*, September 25, 2012b. As of June 3, 2016: http://barentsobserver.com/en/security/russia-sends-mig-31-interceptors-arctic-25-09

———, "Norway and Russia Join Forces in Arctic Response Drill," *Barents Observer*, March 10, 2015. As of June 7, 2016: http://barentsobserver.com/en/security/2015/03/norway-and-russia-join-forces-arctic-response-drill-10-03

Pezard, Stephanie, and Abbie Tingstad, "Keep it Chill in the Arctic," *U.S. News and World Report*, commentary, April 27, 2016. As of June 3, 2016: http://www.usnews.com/opinion/articles/2016-04-27/will-the-arctic-remain-a-warm-spot-in-chilly-russia-us-relations

Putnam, Robert, "Diplomacy and Domestic Politics: The Logic of Two-Level Games," *International Organization*, Vol. 42, No. 3, Summer 1988, pp. 427–460.

Raeste, Juha-Pekka, "HS-gallup: Enemmistö suomalaisista vastustaa yhä Nato-jäsenyyttä [The Majority of Finns Still Oppose NATO Membership]," *Helsingin Sanomat*, March 5, 2015. As of June 7, 2016:
http://www.hs.fi/kotimaa/a1425450355649

Rahbek-Clemmensen, John, "Carving up the Arctic: The Continental Shelf Process Between International Law and Geopolitics," *Arctic Yearbook 2015*, Northern Research Forum and the University of the Arctic Thematic Network on Geopolitics and Security, 2015.

Rees, Andrew, and David Sharp, *Drilling in Extreme Environments: Challenges and Implications for the Energy Insurance Industry*, London: Lloyd's, 2011. As of May 16, 2016:
https://www.lloyds.com/~/media/lloyds/reports/emerging%20risk%20reports/lloyds%20drilling%20in%20extreme%20environments%20final3.pdf

Resolution on Arctic Governance, Brussels, Belgium: European Parliament, October 9, 2008. As of June 3, 2016:
http://www.europarl.europa.eu/sides/getDoc.do?type=TA&language=EN&reference=P6-TA-2008-474

"Russia Launches Military Drills in the Arctic," Agence France-Presse, August 24, 2015.

"Russia Prepares for Arctic Terrorism," *Maritime Executive Newsletter Online*, December 31, 2015.

"Russia Sees Arctic as Naval Priority in New Doctrine," BBC, July 27, 2015. As of June 3, 2016:
http://www.bbc.com/news/world-europe-33673191

Schelling, Thomas, *Strategy of Conflict*, Cambridge, Mass.: Harvard University Press, 1963.

Scudellari, Megan, "An Unrecognizable Arctic," National Aeronautics and Space Administration, July 25, 2013. As of June 14, 2016:
http://climate.nasa.gov/news/958/

Sergunin, Alexander, and Valery Konyshev, "Russian Military Activities in the Arctic: Myth and Realities," *Arctic Yearbook 2015*, Northern Research Forum and the University of the Arctic Thematic Network on Geopolitics and Security, 2015, pp. 404–407.

Sharkov, Damien, "NATO and Russia 'Preparing for Conflict,' Warns Report," *Newsweek*, August 12, 2015. As of July 21, 2016:
http://europe.newsweek.com/nato-russia-preparing-conflict-warns-report-331499?rx=us

Smith, Laurence C., and Scott R. Stephenson, "New Trans-Arctic Shipping Routes Navigable by Midcentury," *Proceedings of the National Academy of Sciences of the United States of America*, Vol. 110, No. 13, March 26, 2013, pp. E1191–E1195. As of January 18, 2017:
http://www.pnas.org/content/110/13/E1191/1.full

Smith-Windsor, Brooke A., *Putting the 'N' Back into NATO: A High North Policy Framework for the Atlantic Alliance?* NATO Defense College Research Division, No. 94, July 2013.

Solli, Erik, Elana Wilson Rowe, and Wrenn Yennie Lindgren, "Coming into the Cold: Asia's Arctic Interests," *Polar Geography*, Vol. 36, No. 4, 2013.

Staalesen, Atle, "Opening the Northern Sea Route Administration," *Barents Observer*, March 21, 2013. As of June 3, 2016:
http://barentsobserver.com/en/arctic/2013/03/opening-northern-sea-route-administration-21-03

———, "New Reality for Norwegian Defence," *Barents Observer*, April 30, 2015. As of June 7, 2016:
http://barentsobserver.com/en/security/2015/04/new-reality-norwegian-defence-30-04

———, "Crisis-Ridden Government Cuts Money for Icebreakers," *Barents Observer*, March 16, 2016. As of June 7, 2016:
http://thebarentsobserver.com/industry/2016/03/crisis-ridden-government-cuts-money-icebreakers

Staun, Jørgen, *Russia's Strategy in the Arctic*, Copenhagen, Denmark: Royal Danish Defence College, March 2015.

Stephenson, Scott R., and Laurence C. Smith, "Influence of Climate Model Variability on Projected Arctic Shipping Futures," *Earth's Future*, Vol. 3, No. 1, 2015, pp. 331–343.

Stephenson, Scott R., Laurence C. Smith, Lawson W. Brigham, and John A. Agnew, "Projected 21st-Century Changes to Arctic Marine Access," *Climatic Change*, Vol. 118, No. 3, June 2013, pp. 885–901.

"Стратегия развития Арктической зоны Российской Федерации и обеспечения национальной безопасности на период до 2020 года [Strategy for Development of the Arctic Zone of the Russian Federation and Ensuring National Security for the Period Until 2020]," Moscow, Russia: Russian Federation, February 2013. As of June 3, 2016:
http://government.ru/media/files/2RpSA3sctElhAGn4RN9dHrtzk0A3wZm8.pdf

Stroeve, Julienne, Mark Serreze, Sheldon Drobot, Shari Gearheard, Marika Holland, James Maslanik, Walter Meier, and Theodore Scambos, "Arctic Sea Ice Extent Plummets in 2007," *EOS*, Vol. 89, No. 2, January 8, 2008. As of January 18, 2017:
https://www.researchgate.net/profile/W_Meier/publication/248820158_Arctic_Sea_Ice_Extent_Plummets_in_2007/links/0deec535e81c7e2f09000000.pdf

Stromquist, Emily, and Robert Johnston, *Opportunities and Challenges for Arctic Oil and Gas Developments*, Washington, D.C.: The Wilson Center, 2014. As of May 12, 2016:
https://www.wilsoncenter.org/sites/default/files/Artic%20Report_F.pdf

Supreme Headquarters Allied Powers Europe Public Affairs Office, "Finland and Sweden Sign Memorandum of Understanding with NATO," NATO, September 5, 2014. As of June 7, 2016:
http://www.aco.nato.int/finland-and-sweden-signing-a-memorandum-of-understanding-with-nato-for-operational-and-logistic-support.aspx

Swedish Ministry of Defense, *NORDEFCO Annual Report 2015*, January 2016. As of May 12, 2016:
http://www.government.se/globalassets/regeringen/dokument/forsvarsdepartementet/rapporter/nordefco-annual-report-2015_webb.pdf

Tamnes, Rolf, and Sven G. Holtsmark, "The Geopolitics of the Arctic in Historical Perspective," in Rolf Tamnes and Kristine Offerdal, eds., *Geopolitics and Security in the Arctic: Regional Dynamics in a Global World*, Oxon, New York: Routledge, 2014.

Tharoor, Ishaan, "The Arctic is Russia's Mecca, Says Top Moscow Official," *Washington Post*, April 20, 2015. As of June 7, 2016:
https://www.washingtonpost.com/news/worldviews/wp/2015/04/20/the-arctic-is-russias-mecca-says-top-moscow-official/

UNCLOS—*See* United Nations Convention on the Law of the Sea.

United Nations Convention on the Law of the Sea, Montego Bay, Jamaica, December 10, 1982. As of June 15, 2016:
http://www.un.org/depts/los/convention_agreements/texts/unclos/part6.htm

United Nations, Oceans and Law of the Sea, "Submissions, Through the Secretary-General of the United Nations, to the Commission on the Limits of the Continental Shelf, Pursuant to Article 76, Paragraph 8, of the United Nations Convention on the Law of the Sea of 10 December 1982," United Nations Division for Ocean Affairs and the Law of the Sea, website, October 28, 2016. As of January 3, 2017:
http://www.un.org/depts/los/clcs_new/commission_submissions.htm

USGS—*See* U.S. Geological Survey.

U.S. Geological Survey, "Circum-Arctic Resource Appraisal: Estimates of Undiscovered Oil and Gas North of the Arctic Circle," Washington, D.C., USGS Fact Sheet 2008-3049, 2008. As of June 29, 2016:
https://pubs.usgs.gov/fs/2008/3049/fs2008-3049.pdf

Ven Bruusgaard, Kristin, "Crimea and Russia's Strategic Overhaul," *Parameters*, Vol. 44, No. 3, Fall 2014, pp. 81–90.

Walker, Shaun, and Sam Jones, "Arctic 30: Russia Changes Piracy Charges to Hooliganism," *The Guardian*, October 23, 2013. As of June 30, 2016:
http://www.theguardian.com/environment/2013/oct/23/arctic-30-russia-charges-greenpeace

Walton, Marsha, "Countries in Tug-of-War over Arctic Resources," CNN, January 2, 2009. As of June 3, 2016:
http://www.cnn.com/2009/TECH/science/01/02/arctic.rights.dispute/index.html?eref=rss_tech

Weitz, Richard, "Russia's Defense Industry: Breakthrough or Breakdown?" International Relations and Security Network, March 6, 2015. As of January 23, 2017:
http://www.hudson.org/research/11322-russia-s-defense-industry-breakthrough-or-breakdown-

White House, *National Strategy for the Arctic Region*, Washington, D.C., May 10, 2013.

———, *Implementation Plan for the National Strategy for the Arctic Region*, Washington, D.C., January 2014.

Wilson Rowe, Elana, and Helge Blakkisrud, *Great Power, Arctic Power: Russia's Engagement in the High North*, Policy Brief, Norwegian Institute of International Affairs (NUPI), February 2012.

———, "A New Kind of Arctic Power? Russia's Policy Discourses and Diplomatic Practices in the Circumpolar North," *Geopolitics*, Vol. 19, No. 1, 2014, pp. 66–85.

Winger, Gregory, and Gustav Petursson, "Return to Keflavik Station: Iceland's Cold War Legacy Reappraised," *Foreign Affairs*, February 24, 2016.

Young, Oran R., "Whither the Arctic? Conflict or Cooperation in the Circumpolar North," *Polar Record*, Vol. 45, No. 232, 2009, pp. 73–82.

Zysk, Katarzyna, "Russia's Arctic Strategy: Ambitions and Constraints," *Joint Force Quarterly*, No. 57, second quarter, 2010.

———, "The Evolving Arctic Security Environment: An Assessment," in Stephen Blank, ed., *Russia in the Arctic*, Carlisle, Pa.: U.S. Army War College, July 2011, pp. 91–138.

———, "Russia and the Arctic: 'Territory of Dialogue' and Militarization," briefing presented at the Arctic Frontiers Conference, Tromsø, Norway, January 29, 2016.

Zysk, Katarzyna, and David Titley, "Signals, Noise, and Swans in Today's Arctic," *SAIS Review of International Affairs*, Vol. 35, No. 1, Winter–Spring 2015, pp. 169–181.